Flash Floods in Vietnam

Le Huy Ba • Thai Van Nam • Le Hung

Flash Floods in Vietnam

Causes, Impacts, and Solutions

 Springer

Le Huy Ba
Environment, Resource and Climate Change
Ho Chi Minh City University
of Food Industry
Ho Chi Minh City, Vietnam

Thai Van Nam
Vice Director of Applied Sciences Institute
of HUTECH
Ho Chi Minh City University of Technology
Ho Chi Minh City, Vietnam

Le Hung
Head of HANE Foreign Relations
Department
Association for Conservation of Nature
and Environment
Ho Chi Minh City, Vietnam

ISBN 978-3-031-10534-0   ISBN 978-3-031-10532-6   (eBook)
https://doi.org/10.1007/978-3-031-10532-6

This Springer imprint is published by the registered company Springer Nature Switzerland AG
The registered company address is: Gewerbestrasse 11, 6330 Cham, Switzerland

Preface

Humanity is increasingly facing natural hazards due to the effects of global climate change. Among the natural disasters that cause serious damage to humans and property are land degradation, floods, desertification, erosion, and mountain, riverbank and coast landslides, causing the loss of land, fields, and villages that have been settled for hundreds of years. A flood is a natural phenomenon that has occurred more frequently in recent decades and is a constant threat to people everywhere in the world. In the mountainous provinces, yearly flash floods also cause a loss of hundreds billions VNĐ in damage, devastate the environment and ecosystems, kill, and injure many people.

Flash floods usually appear in tropical river basins that are affected by monsoons and storms. Floods in Asia, particularly in Vietnam, are increasing in both frequency and intensity. Flash floods and landslides are actually unpredictable, but their damage can be limited by "active prevention." Because flash floods are immediate and unforecastable, for the preventive measures proposal, we need to identify and localize areas prone to flash floods. The most appropriate solutions will be built depending on the characteristics of topography, geomorphology, soil, and population distribution.

The irony paradox is that normal – seasonal – floods are familiar floods that have brought a source of fresh water, alluvium, aquatic organisms, and diverse ecosystems to the Mekong Delta for many years, increasingly scarce because the Mekong River water source is exhausted for all kinds of reasons from upstream (of which mainly due to dams blocking the river in China and Laos), but flash–tubing floods (the most harmful partial floods that everybody is tired and afraid of) are appearing increasingly massive and destructive in many large mountainous and midland regions of Vietnam.

Since 2020, from the beginning of July to the end of October, floods and rains have occurred extremely complicated due to the full influence of East Sea storms on the Northwest, North Central, and Central Highlands provinces that caused water from rivers and streams to rise and flow down low basins, causing flash and tubing floods and large landslides, and seriously damaged the people's lives, houses, works, bridges, roads, and property, and stagnated production, upset daily life, and paralyzed traffic. Flood disasters have happened and are happening in these places with tremendous devastation, spreading tragedy everywhere, causing extremely difficult consequences.

Remember that flash floods occur not only in the highlands but also during highly concentrated rainfall. The premise factor for flash floods is soil erosion. When flash floods appear, they affect the status of soil erosion, with the surface peeling away as an increasing process of flash floods harmful effects. In addition, the flood flow also increases the topsoil erosion many times. These two processes are closely related; sometimes they merge into one "flash flood–erosion" process with two reciprocal effects. In addition to, or parallel to these two processes, the landslides on mountain-sides, cliffs, and on stream or river sides, even coast-lines have the resonance of erosion, flood, rain, and flow.

From these theoretical studies and experiments, combined with field investigation, we would like to show 15-years of results with some extracts and references from scientific reports of domestic and foreign colleagues. Topics include soil erosion implemented in the basin of Dong Nai river, Serepok river, Dak Mo, Vinh An sub-basin regions; flash floods in six mountainous districts of Quang Ngai, Binh Phuoc, Dak Lak provinces, landslides occurring on Dong Nai river, mountainous areas in the West of Quang Nam, Quang Ngai; Tien river, Ben Tre river; coast erosion observed, recorded, and referenced in coastal provinces from Binh Thuan to Kien Giang. This is also a reference document for Lecturers and Students at Universities.

Because the mechanism characteristics of the flash flood are absolutely identical in all places where the conditions of terrain are narrow, steep, and the amount of water (from rain or snow) converges much and quickly, this scenario progress of flash flood occurs in almost all places on this planet with the same topographic, hydrological, and meteorological conditions for flash flood risk. Not only that, other problems of monitoring, forecasting, early warning, and solutions of prevention, response, treatment-emergency or radical anywhere in the world are similar (only depending on resources of different places) for this kind of disaster. This implies that although the study is based on Vietnamese areas, it has an international cognitive background and application.

We would like to thank the local units and people for facilitating research funding and providing data: Department of Science and Technology, of Natural Resources and Environment, of Agriculture and Rural Development, Sub-Departments of Hydrology, Water Resources, of Environment and People of Binh Phuoc, Quang Ngai, Dak Lak, Kien Giang, Bac Lieu, and Ca Mau provinces. We also express our gratitude to international colleagues from Belgium, Denmark, the UK, Germany, Japan, Thailand, Indonesia, and Malaysia who have helped our research.

Ho Chi Minh City, Vietnam   Le Huy Ba

Contents

About the Authors

D. Sc Le Huy Ba Head of Institute of Environment, Resource and Climate Change, HCMC University of Food Industry (HUFI), Vietnam

Prof. D Sc Le Huy Ba successfully defended his PhD thesis (level 1), and was nominated for a second degree and successfully defended his Doctor of Science Thesis in December 1988, in Sofia, Bulgaria. He is Prof. Sci. Dr. experienced in Environmental Ecology, Soil Ecology, Water Environment, especially about Environmental Toxicology. He determined the role of peroxidase enzyme in rice leaf, which is toxic-resistant. He discovered the change and actions of heavy metal toxins in soil, water, and sediments and their influence on human health and rice plants. Moreover, he determined the condition of lateralization in highland, sulfate acidization in Mekong Delta and rule of dynamic and the effect of toxic ions in the environment on plants and humans. He is a Member of the Biology Council for National Professor Title of Viet Nam. He has published 32 books, among them, there are 7 monography books. He has won 3 National Book Awards. He is Viet Nam's Elite House with 52 years teaching, 21 years is full professor.

Thai Van Nam Ho Chi Minh City University of Technology, Ho Chi Minh City, Vietnam

Vice Director of Applied Sciences Institute of HUTECH (University of Technology)

Dr. Van Nam Thai is an Associate Professor at HUTECH Institute of Applied Sciences (HIAS), Ho Chi Minh City University of Technology – HUTECH. He currently works as Vice Director of HIAS, a Senior Lecturer in the field of Sustainable Energy, Environmental Engineering and Management. He obtained his master degree at Vietnam National University Ho Chi Minh City (VNUHCM), Vietnam – 2003 and doctorate degrees at Osaka University (Handai), Japan – 2011.

Dr. Thai majored in Environmental Management Systems, Environmental Engineering, Water and Soil Resources Engineering, Environmental and Health Risk Assessment. His skills and expertise are in Health and Environmental Risk Assessment, Life-Cycle Assessment, Environmental Analysis, Environmental Management, Sustainable Consumption and Production, Environmental Modeling, Land Degradation and Soil Remediation. He has published over 100 research items, i.e., 67 articles (36 ISI papers), 17 books and book chapters, and 22 conference papers. One of his publications is in the Journal of Cleaner Production with Citescore: 10.9, IF: 9.279, ISI-Q1.

Le Hung, MS/MBA Association for Conservation of Nature and Environment (HANE), Ho Chi Minh City, Vietnam

Head of HANE Foreign Relations Department/ Deputy Editor of ENVIRONMENT WORLD Magazine/Manager of R & D Department of Renewable Energy RETECK VN

In 1978, after graduating from the College of Military Economics, Le Hung specialized in Planning and Cartography, serving the construction of military farms and economic zones. In 1982, he earned a Master of Environmental Economics from the National Economics University and moved to Saigon Port to be in charge of maritime services and concurrent monitoring of pollution for marine environmental (Marpol). Later, he completed the MBA certificate (1998–2000) in Business Administration by University of Montreal – Quebec CANADA in association with the HCMC University of Economics.

In 2017, he was in charge of External Relations at the Association for Conservation of Nature – Environment in Ho Chi Minh City, and Deputy Secretary of the World Environment Online Magazine with consulting, communication, and lecturer works on environmental issues and the circular economy. He has many specialized articles published in Vietnam's Science & Technology and Natural Resources & Environment newspaper and most recently (2019–20,220) edited, translated, and wrote Environmental Science for the author group chaired by Prof. Dr. Le Huy Ba and also had three articles in International Environmental Journal and World Water Resources report.

Abbreviations

A&RD	Agriculture & Rural Development
AFC	Agricultural – Forestier Combination
CEC	Cation Exchange Capacity
COD	Chemical Oxygen Demand
DEM	Digital Elevation Model
EC	Electrical conductivity
ES	Ecologic System
ES	Erosion Status
FAO	Food & Alimentation Organization
FF	Flash Flood
GIS	Geographic Information System
LUS	Land Use Status
MRD	Mekong River Delta
MRE	Ministry of Resources and Environment
NS	Natural Surface
OC/H	Organic Content/Humus
PE	Potential Erosion
R&E	Resources & Environment
RS	Remote Searching
SALT	Sloping Agricultural Land Technology
TDS	Total Dissolved Solid
TF	Tubing Flood
TSS	Total Suspended Solid
VS	Vietnam Standard
VSzn	Vietnam Standardization

Chapter 1
Overview of Flash Floods

Abstract This chapter presents an overview of the most destructive natural disasters in the world – flash floods, which cause 5000 deaths globally every year and 50% of damage to property, infrastructure, and the agriculture industry. The lethal nature of flash floods is attributed to the fact that they occur on a small spatial scale within a short time, quickly creating surface flows in all suitable terrains and weather, appearing anywhere on this planet. Learning about flash floods and proposing solutions to prevent and limit their harm is necessary and urgent.

The purpose of this study was to successfully build a zoning map and a flash flood risk warning model at high precision, together with further research, applying AI in the future to accurately forecast natural disasters in real time. Through the case study of Dak Lak province in Vietnam, the study also assessed the impact of flash floods on the environment and socio-economy, and then presented a case study of Binh Phuoc province to outline how to prevent and overcome disasters.

Keywords Flash flood lethal nature · Flash flood impacts · Fluctuation rate · Xí Mân and Hoang Su Phi

1.1 Introduction

Among the world's most destructive natural disasters, flash floods are ranked first with more than 5000 deaths annually globally and a mortality rate more than 4 times greater than other types of flooding. In all flood-related hazards, flash floods account for 80% of deaths and 50% of loss to property, infrastructure, and the agriculture industry.

All terrains and weather conditions are conducive to flash floods, and they appear indiscriminately anywhere on this planet. Flash floods are typically associated with short, high-intensity rainstorms; therefore, their characteristics cause the very short response time and the potential to severely impact and damage communities worldwide. To date, despite their scientific and social importance, the fundamental processes of a flash-flood reaction are not clearly understood. The newly enriched study of flash floods is providing useful contributions that aims to provide a review of the hydrological mechanisms driving hill-slope runoff response to intense rainfall and to characterize the runoff evolution from the extreme flash floods.

The lethal nature of flash floods is largely attributed to its characteristics: events occurring on small areas in very short time under conditions of quick formation of surface runoff (for example, intense rain or precipitation on highly saturated soils in mountainous terrain). The specialists define flash flooding events as flood events which rise and fall rapidly with little or no advanced warning, usually as the result of intense rainfall over a relatively small area.

In addition to heavy precipitation, the meteorological, climatic, and physiographic influences on flood occurrence include the precipitation intensity, topography, and soil characteristics. The significance of these factors in comparison to the climatology of heavy precipitation with flash flood occurrence leads to the conclusion that flash floods occurred many times less frequently than heavy precipitation.

Several studies have examined changes in larger scale (>1000 km^2) flooding events under climate change and the local scale impacts of sea level rise on coastal areas. Other studies have examined changes in atmospheric conditions related to the increased flood risk to infer impacts of climate change scenarios on flood risk.

Based on research of the Intergovernmental Panel on Climate Change (IPCC), an anticipated impact of climate change is regional variation in precipitation magnitude and variability of rainfall extremes, and flooding occurrence toward an increased risk of extreme phenomenon. Changes in precipitation and temperature have already been observed and attributed to human influence. Major impacts are changes in temperature, snowfall, snowmelt, and sea level rise.

The impacts of flash floods on water resources may be substantial, and climate change may play a role. The major impacts of climate change focus on changes in temperature, precipitation, and sea level rise, as well as the impact examples such as potential increases in wildfires and energy demand.

A study of twentieth century changes in climatic indices, including the Pacific Decadal Oscillation (PDO) and El Nino Southern Oscillation (ENSO), and associated increases in flood risk from climate change and in terms of atmospheric mechanisms associated with flooding, the "atmospheric stream" phenomenon, concluded that storms associated with the warm-wet atmospheric river conditions may increase in the future and thus there would be an increased flood risk compared to the historical record.

Few studies have explored the potential climatic change impacts on small scale flooding events. Precipitation estimation with high resolution for smaller scale hydrologic impacts have required probabilistic calculations or dynamic downscaling methods to estimate the rainfall at daily sub-time scales or spatial resolution of a few 10 s of 1 km. The large-scale climate model simulations, when brought to appropriate scales using dynamic downscaling methods, can provide useful information on small-scale flash flooding in a region.

The overall motivation was to develop a methodology that could examine the spatial character and variability of flash flood occurrence, over a large region (about 1 million ha) but with high spatial detail (about 1000 ha). The methodology synthesizes principles and practices from the fields of meteorological modeling and hydro-geomorphology.

1.2   The Necessity of Flash Flood Knowledge

Flash floods are extreme hydro-meteorological events. However, at this time, modeling efforts remain the only approach feasible to examine the climatic scale variability and change of flash flood occurrence over several decades. Improved understanding of such flash floods and the hydrologic response to climate variability and change is an important and necessary advancement.

Therefore, it is imperative to research flash floods then propose solutions to prevent and limit their harm. The research results will be an essential overview for everyone from citizens to researchers and managers.

For the people, they will know where they are and how dangerous it is, if they have or do not have the disasters, landslides, flash floods, etc., to make a decision to stay there or not, and if so, what to do in that case.

For administrators and researchers, they will have an overview of flash floods and the risks of their occurrence frequency and extent of their destruction, with the aim to guide the appropriate planning of residential areas, urban areas, or evacuation from dangerous zones and have long-term and sustainable socio-economic development goals; also giving suitable solutions to minimize consequences of natural disasters.

1.3   Objective of the Study

The objective of this research is to apply the model of simulation method combined with geographic information system (GIS) and remote sensing (RS) technology to successfully build the zoning map and warning model of flash flood risk, which will be better than the used traditional maps and models. The model is also used to evaluate the performance and accuracy of the results. This is the scientific basis for further research to apply artificial intelligence (AI) in the near future for accurate prediction of natural disasters in real time.

The purposed target of flash flood forecasting map is to (1) Instruct residents and managers to clearly find the area affected by different types of flash floods, the process nature of the formation and development, the intensity and probability of flash flood occurrence; (2) Orientation of material reserve, prevention strategy, and rescue plan when necessary; (3) Make strategic plans for investment in facilities and disaster prevention solutions for all management levels; (4) For the improvement of safety and enforcement in project planning and sustainable development of territorial exploitation.

The ultimate application orientation of the whole topic is aimed at:

- Assessing the real status of effects, the impacts of tubing, flash floods, landslides in urban areas, rural population spots in the Tay Bac mountainous area, Central Highlands, and Central Coast.
- Reviewing the urban planning, population spots in hilly, mountainous, and coastal areas at risk of flash floods and landslides.
- Developing the effective structural, non-structural solutions, and monitoring management systems for flash flood disaster prevention.
- Becoming an experience lesson for many other countries and regions in the world to economize the time & effort of further investigation and eliminate painful experiments.

In fact, from the topic and research project, we obtain the essential cores to understand and avoid the harmful effects of flash floods, to help to warn of their risk in the mountainous and highland provinces and key areas at high vulnerability. The study took the three Vietnamese provinces (Dak Lak, Bình Phước, and Quảng Ngãi) where flash floods often occur as "case studies," from there, preventive measures were proposed to minimize damage and serve as a way to plan socio-economic development in the risky areas.

1.4 Contents of the Study

- Knowledge of flash floods and related problems
- Current status of flash floods in the world and Vietnam
- Causes of flash flood formation
- Stages of flash flood formation
- Basic features of flash floods
- Criteria to identify flash floods
- Assessment of environment and socio-economy impacts
- Signs for flash flood prediction
- Socio-economic characteristics related to flash floods (case study: Dak Lak)
- Database establishment for the map of flash flood risk (case study: Dak Lak)
- Map of flash flood risk (case study: Dak Lak)
- Preventing and overcoming flash floods (case study: Bình Phuoc Province)

1.5 Application Range

The mountainous provinces of North Vietnam, the Central Highlands, the mountainous Central region, as well as all countries in the world under condition of concentrated heavy rainfall or snow, and melting ice are at risk because of the identical characteristics of the flash flood mechanism (high water intake, short duration, and narrow slope terrain).

1.6   Benefits of Flash Flood Research

This research hopes to limit the socio-economic damage to at risk provinces (Dak Lak, Quang Ngai, and Binh Phuoc), contribute to stabilizing the lives of those living in the communities in areas prone to flash flood risk, prevent and control natural disasters, reduce disadvantage effects on the regional environment, and protect the ecology.

Research results of this topic are an essential overview for any people as well as managers and researchers. This is the basis for population distribution, productive structure transformation, infrastructure construction, especially for the disaster management and prevention, forecasting, researching, and warning of the expected occurrence of floods.

1.7   Situation of Flash Flood Research

• **Overseas research**

There have been many research topics and reports on flash floods around the world, typically summarized as:

– Topics of WHO
– WMO studies, reports and avoidance guidelines
– Reports of World and United Nations Environmental Organizations: UNEP, WWF, GEF, IUCN, World Bank, UNDP
– Research summaries, reports of governmental & non-governmental organizations: Academia, ResearchGate, NSSL, ScienceDirect, AMS American Meteorological Community, Pergamon Environmental Hazards (USA) on extreme weather, OCHA ReliefWeb, Swiss Frontiers, EGU EU geo-scientific information, UN Global Compact's MDPI – United Nations, EUROPE PMC
– University faculties of natural sciences of Great Britain, France, Germany, Russia, the Netherlands, Italy ... and many Asian countries. It can be said that this research topic is available in most countries in the world at different levels.
– Scientific journals: La recherche, Science et Vie, Euro Scientist, New Scientist, BBC Focus, Australian Science, Current Science, Focus, Vokrug Sveta, Popular Science, Discover, Science News, Springer, MIT Technology Review, The scientist.v.v.

• **Research situation in Vietnam**

Vietnam, similar to other countries of the world, suffers extensive damage from storms, floods, landslides, and flash floods. According to statistics, in Vietnam flash floods occur almost annually, at an increasing pace.

From 1990 to 2020, 500 flash floods occurred in mountainous areas nationwide; flash floods left 965 people dead and missing, 628 people injured, 23,280 houses swept away, 114,849 houses flooded and damaged, 197,879 ha of rice, other plant crops, and many transport and irrigation systems were ruined. Total material cost of the catastrophes is estimated at about 21,915 billion VND. The big flash floods caused extensive damage to people, houses, and infrastructure. In recent years, in the Central Highlands region, flash floods have increased in frequency. According to statistics, in the past 15 years, there have been 18 large and particularly large normal floods and flash floods causing widespread flooding.

In August 2005 alone, at least five local flash floods occurred in the province of Kon Tum and killed three people; about 1500 ha of food and plant crops were lost, and 31 temporary dams were swept away, five cable-stayed bridges, 22 major irrigation works, three schools, six concrete bridges and tens of kilometers of national highways and provincial roads were badly damaged. The estimated total damage was nearly 15 billion. The localities that suffered the most serious damage were the semi-submerged area of Ya Ly lakebed of the Sa Thay district, Dak Ha district, Dak Glei, and Dak Tô.

Listed below are the most recent research topics on natural disaster and flash flood risk, made by Vietnamese scientists.

– The State-grade project KC.08 *"Research & development of zoning map of Vietnam regions in natural environmental risks"* by chief author Prof. Dr. Nguyen Trong Yem was completed in 2006. The research team established two maps: landslide map in national caliber, flash flood map, rock-mud floods map at scales of 1:100,000 and 1:500,000. This is the first time that Vietnam has established these maps and they are an important basis for scientific research as well as the orientation for further research.

These maps warn of the danger of flash floods and landslides in key areas: the east side of Hoang Lien Son mountains, the right bank of the Red River, including the districts of Bat Xat, Sa Pa, and Lao Cai city; Nâm Lay river basin in the Mương Cha district and Nâm Rom river basin, including part of Dien Bien east, west districts and Dien Bien city; the upstream area of the Gam river in Yen Minh district, tributaries on the left bank of the Chay river and the right bank of the Lô river in districts Xí Mân and Hoang Su Phi (Ha Giang).

– In 2007, the provincial-grade project *"Development of risk maps and feasible measures to prevent the harmful effects of tubing – flash floods in Quang Ngai province"* was chaired by chief author Prof. Dr. Le Huy Ba. According to the results from the flash flood hazard zoning map, the research team concludes on the flash flood risk in six mountainous districts of Quang Ngai province with high slope (over 25° steep angle) as follows:

• **High risk areas (level 1)** account for 4.87%, comprising: Bao mountain area (adjacent to Tra Xinh, Son Bua, Son Mua, Son Bao communes); Ta Cun mountain (the area adjacent to Tra Hiep and Tra Quân); Ca Dam mountain (the area adjacent to Tra Nham and Tra Tan communes); Ba Ang mountain (the area adjacent to Nghia Son and Son Nham communes); Ha Peo mountain, Rong mountain (the area adjacent to the communes Son Dung, Son Tinh, and Son Lap); Ba Tu mountain (Ky Son commune), Lang Râm mountain (the area adjacent to Ba Le and Ba Nam communes); Pa Xa mountain, Sang mountain (the area adjacent to Tra Son, Tra Lam, and Tra Nham communes).

- *Average risk areas (level 2)* account for 27.39%, comprising: Tang river area (the area adjacent to the communes of Tra Xinh, Tra Tho, and Son Bao), the area of Tang river, the area of Tin mountain, Mount Rin, river Rinh, the area of Ngoc Ven mountain (Son Dung commune), the area of the tributaries of Tra Khuc river (the area adjacent to Son Linh, Son Hai, and Son Cao communes), the Re river area (the area adjacent to Son Ky and Son Ba communes).
- *Low risk areas* account for 17.29%.
- *Very low risk areas* account for 50.45%.

– In 2009, the provincial-grade topic was studied by Prof. Dr. Sc. Le Huy Ba: *"Development of risk maps and feasible measures to prevent and mitigating the flash flood damage in Binh Phuoc province."* The Binh Phuoc province zoning map of flash flood risk showed:

- *Areas in high risk ratio of flash flood* account for only 0.52% of the total, distributed in the northwest of Phuoc Long, northeast and west of Bu Dang and scattered in some places in Dong Phu, Dam fountain basin (downstream of Thac Mo and Can Dơn, communes of Bu Nho, Long Ha, Son Giang, and Thac Mo); Rat fountain basin (Dong Xoai town, Dong Phu); Basin of Cam, Dak Ơ, Can Lê fountains (Binh Long district); Dak Rlap river, Bu Dăng district; DakWar fountain of Bu Dăng district.
- *Areas in average risk ratio of flash flood* account for 7.03% of the total, and are scattered in all districts of Binh Phuoc province, except Chon Thanh district.
- *Areas in low risk ratio of flash flood* account for 92.45% of the total.

In another region – Dak Lak – before the stormy season of every year, Dak Lak province also had a policy of relocation and support for households living in areas prone to flash floods. Therefore, limiting and preventing flash floods in the province is also one of the top concerns of provincial leaders.

– In 2010, the Provincial-grade Scientific Research project: *"Development of risk maps and feasible measures to prevent tubing – flash flood damage in Dak Lak province"* was done by Prof Le Huy Ba & collaborators, with survey results, investigation, synthesis, and analysis. They assessed that Dak Lak province has five areas prone to flash floods. In other words, there are five main areas at very high risk of flash floods:

- **Zone 1:** Mostly includes the Cư M'gar district (area at very high risk of flash flood, accounting for 14.72%); the basin of the streams Ea Tul and Ea Huar, starting from 700 m upstream (Cư Dlie Mnông commune) flowing 200 m downstream (Ea Huar commune), and the flash flood area includes communes: Cư Dlie Mnông, Ea Tar, Ea Kpam, Ea Hđinh, Ea Kiêt, Cư M'Gar, Ea

Mroh, Quang Hiep, and Ea M'nang (belonging to Cư M'Gar district), and Ea Huar and Ea Wer (Buôn Dôn District).

- **Zone 2:** Mostly includes the Krông Năng district (area at very high risk of flash flood, accounting for 12.45%) and the Krông Năng river basin, which flows from 750–800 m in the upstream to 400–500 m downstream, through the communes: Dliê Ya, Cư Klông, Ea Tam, Phu Loc, Tam Giang, Ea Puk, Krông Năng Town, Phu Xuan, Ea Dah belongs to Krông Năng and Xuan Phu districts, Ea Kar town is in Ea Kar district.
- **Zone 3:** includes the M'Drăk district (with 3.98% of the area at very high risk of flash flood) Krông Jing stream. Flowing from the altitude of 500 m in communes of Cư M'Tar and Cư Kroa, falling to a height of 400 m in Krông Jing commune and M'Drăk town.
- **Zone 4:** includes the Lak district (4.98% of the area is at high risk of flash flood) stretching along the streams of Dak Phơi, Dak Liêng in Lăk district. Upstream from 1000–1200 m (Dak Phơi commune, Bông Krang) down to 400–500 m in two communes Dak Nuê and Dak Liêng and a part of Dak Liêng town.
- **Zone 5:** includes the Ea Hleo district (8.80% of the area is at very high risk of flash flood). Communes in Ea Hleo district: Ea Tir, Ea Khal, Cư Amung, Ea Wy, Cu Môk, and Ea Drăng town.

Based on the actual conditions of Dak Lak province – highly fragmented topography, limited vegetation, high annual rainfall, and dense irrigation system – the research team has proposed the following suitable solutions:

- Construct upstream water-regulating works; build solid reservoirs for both irrigation and flood regulation.
- Upgrade and repair the degraded irrigation works.
- Increase the flood drainage capacity of the conductor-bed.
- Schematic land planning in the basin to aid in flash flood avoidance.
- Land use needs to be more reasonable than before.
- Planting more forests to protect better than before.

Chapter 2
Knowledge of Flash Floods and Related Problems

Abstract This chapter provides general knowledge on flash floods, how to distinguish between flash floods and normal floods, the causes ranging from natural to human, and the formation mode of a flash flood into an overwhelming disaster. The chapter describes the current situation of flash floods in Vietnam and in the world.

This chapter highlights the basic properties of flash floods (sudden/intense/dense/chain/accelerated) and distinguishes the different types of flash floods (flooding/side sweep/mud–rock/broken structure/combination) for early identification of flash floods and avoidance. The chapter also assesses the environmental, natural, and socioeconomic impacts caused by flash floods.

Keywords Flash flood classification · Velocity and destructiveness · World flash floods · Flood overlaps

2.1 Concept

2.1.1 General Awareness

A flash flood is known as a flood in which the water amount increases during or several hours after a heavy rainfall. Flash flooding occurs in small basins for a short time. Partly due to the drastically rapid rise of flash floods, the damage can be very severe.

Many hydrological factors are associated with the occurrence of flash floods: topographic slope, soil type, vegetation cover, human habitat, previous precipitation, etc. In steep, rocky terrain or in heavily urbanized areas, even a relatively small amount of rainfall can cause flash floods. These hydrological factors determine the response to precipitation events of the basin. Thus, flash floods are clearly the result of a mixture of both meteorological and hydrological factors.

- The places of flash flood occurrence: mainly in mountainous areas, with steep hills, at the heads of small tributaries.

- Frequently the vegetation cover is completely lost or only retained shrubs, and a few big/old trees.
- There are heavy rains, concentrating over 300 mm rainfall.
- The soil in the whole region or sub-region is too water-saturated, becomes muddy mixed with rocks, but between rocks and rocks or soil no longer linked.
- Some obstacles stop the surface flow for a while, then, due to the pressure (the large kinetic energy of the flow breaks the obstruction – temporary dam), the water flow from the accumulation of maximum potential energy $W(t)$ becomes kinetic energy $W(d) = mv^2/2$.
- From this formula, the larger the flow mass (including rock + soil + water + mud ...), the greater the kinetic energy $W(d)$. Moreover, $W(d)$ also depends on the slope angle. The larger the angle and longer the slope, the stronger $W(d)$; therefore, on the way, it "sweeps away" all and drowns everything.

2.1.2   Distinguishing a Flash Flood from a Normal Flood

By the unifiable definition of experts, a normal flood is known as "water overflowing into normal dry land." Inundation from a normal flood can originate in many places, such as existing waterways. Rivers and fountains can cause flooding in surrounding areas and flooding occurs over a long period, from some days to some weeks.

Normal floods often have a slow, easily controlled, and gently variable progression – even cyclical, with little loss of human life but heavy and profound damage to crops, farming, and structures. There are some seasonal floods that people still look forward to in order to improve their lives, for example, the floating flood in Dong Thap Muoi.

Brief, normal floods are the floods that occur due to spilling water from the large overflow of rivers and streams rising to inundate the whole basin, often lasting for a long time, seasonal and regular.

In contrast, flash floods are generally surprisingly quick. Although floods have complex origins, flash floods often come from immediate heavy rains or other sources, such as a breakage of a dam or a dyke. Flash floods are any sudden flow of water into a confined area, usually within 6 h or less, often characterized by violent vortices, drastic currents filled with solid matter that rips through the bed of rivers, streams, streets or canyons, valleys, and sweep away everything in front of them.

Event witnesses often report that flash floods are accompanied by a rumbling thunder, horrible sound like a bomb explosion, and a black wall when unexpectedly exposed, and among these witnesses, memories of flash floods are terrifying and unforgettable (Fig. 2.1).

A regular flood and flash flood can be compared to a Marathon and a Sprint; it is the difference between something that comes slowly but will surely destroy due to the lingering water and something that comes and goes quickly with rapid but partial destruction. Normal floods often cause heavy rainfall and widespread damage

Fig. 2.1 Flash flood in Central Highland

but few deaths, while flash floods are lightning fast, claiming a lot of lives, but causing less damage to property because there are few infrastructure works in those areas.

With normal floods, it is still possible to prepare and wait for weather and hydrological news to adjust the level of prevention, but for flash floods, it is necessary to take action at-once to escape immediately after receiving the warning, not waiting for a minute more!

2.2 Description

Flash flood is a type of partial flood in a narrow space such as in mountainous areas or other areas (even cities, towns) with many rivers, streams, and water flows. Flash floods are usually large floods that appear suddenly in a short time in a narrow space (only a few hundred, several tens or a few km²) but have great destructive power due to the rapidly accumulating kinetic energy. Flash floods have intensity, velocity, concentrated flow in the form of drainage pipes and very high water level amplitude, rise quickly and descend quickly, the water flow has a high solids content and can cause tremendous devastation.

The enormous destructive power of flash floods is due to their rapid and resonant operation. When a flash flood forms and moves, the obstructions in front of the surface flow could be suspended several times for a while at first, but then, due to the impulse from kinetic energy, the current potential increases. The currents accumulate enormous potential with the flow volume carrying enough rocks, soil, mud, orphan trees, etc., making the resonance that turns into kinetic energy, and the velocity and destructiveness of flash floods will vary depending on the terrain.

In steep terrain or even in densely urbanized areas, even a relatively small amount of concentrated rainfall can cause flash floods. Narrow areas with few flood divide lines, many slopes, high steep angles, and longer slopes have stronger and more aggressive flash flood, which "sweeps" away all and sinks everything. Because flash floods are the movement phenomenon of a giant water mass from high to low with increasing speed and destructive power, it will cause extremely serious damage to the places it passes through. Despite being strong and destructive, flash floods usually occur quickly, do not happen for longer than 6 or 7 h, and the water recedes after 10–18 h .

2.3 Flash Floods in the World and Vietnam

2.3.1 World Flash Floods

Flash floods are "specialties" of abnormal weather patterns in mountainous or semi-mountainous areas. This terrain style is found all over the world, and therefore many places have suffered this form of devastating natural calamity. Some examples of the most memorable recent flash floods on the planet are as follows:

Flash flood is an environmental risk and sometimes an environmental catastrophe, for example, in China 2020 on the Yang Tze river basin. Our world is facing major changes in nature. Many countries are suffering the dreadful consequences of climate change. Flash floods are one of the natural disasters that cause tremendous damage to human life and property.

Some recent years ago, the phenomenon of flash floods began to accelerate in most countries worldwide, especially in river basins in the tropics affected by monsoon climate and storms, causing heavy losses. The places most often hit by flash floods in old continent Europe are Southern France, Northern Italy, and the Karpat mountains (Fig. 2.2).

The river basins around the San-Gabriel Mount in California State are the most vulnerable to flash floods in the United States. Flash floods occur on the slopes of the Andes Mountains in Mexico, Colombia, Ecuador, Peru, and Chile, in African countries, Australia, and the mountainous basins of the coasts of the Pacific, Atlantic, and Indian oceans. In Asia, flash floods often occur in tropical and subtropical areas such as India, China, Pakistan, Thailand, Nepal, Indonesia, Philippines, Malaysia, Japan, and Vietnam. Flash floods during Asian tropical monsoon and cyclone climates are increasing in both frequency and intensity.

There is a list of typical flash floods all over the world in twentieth and twenty-first centuries to show us how widespread this type of disaster is:

- 1903: Oregon, United States: 247 dead.
- 1938: Los Angeles, California, USA: 115 dead

Fig. 2.2   Flash flood destroyed 1 village in Indonesia

- 1938: Mahia Peninsula, New Zealand: 21 dead
- 1952: Lynmouth, England: 34 dead
- 1963: Petra, Jordan: 23 dead
- 1963: Vajont dam disaster, Italy: 1910 dead
- 1967: Lisbon, Portugal: 464 dead
- 1969: Virginia, USA: 123 dead
- 1971: Kuala Lumpur, Malaysia: 32 dead
- 1972: South Dakota, USA: 238 dead
- 1976: Colorado, USA: 143 dead
- 1997: Arizona, USA: 11 dead
- 2003: Bukit Lawang, Indonesia: 239 dead
- 2006: Surat, Gujarat, India: 100 dead
- 2006: Jember Regency, Indonesia: 59 dead
- 2007: Sudan: 64 dead
- 2009: Manila, Philippines: more than 100 dead
- 2009: Messina, Italy: 37 dead
- 2010: Madeira archipelago: 42 dead
- 2011: Queensland, Australia: 21 dead
- 2011: Philippines: 1200 dead
- 2012: Nepal: 72 dead
- 2012: Krasnodar, Russia: 172 dead
- 2013: Sardinia, Italy: 18 dead, 3000 homeless
- 2013: Port Louis, Mauritius island: 11 dead
- 2013: Argentina: more than 99 dead
- 2013: Uttarakhand, India: 5000 dead
- 2014: Jammu & Kashmir, India: 300 dead

- 2014: Serbia, Bosnia and Croatia: 30 dead
- 2015: Texas, USA: 25 dead
- 2016: West Virginia, USA: 24 dead
- 2016: Maryland, USA: 2 dead
- 2016: South Louisiana, USA: 13 dead
- 2016: Garut Regency. Indonesia: 33 dead
- 2017: West Attiki, Greece: 23 dead
- 2018: Tzafit Canyon, Israel: 10 dead
- 2018: August 20, Raganello Canyon, Italy flood: 10 dead
- 2018: Some parts of Spain, Italy: 15 dead
- 2018: Aude region, France: 15 dead and 99 injured
- 2018: Dead Sea, Israel: 21 dead
- 2019: Southern Iran: 23 dead
- 2019: Northern Afghanistan: 16 dead

In addition to the floods listed above, there are several other impressive floods – these are the flash floods accompanying normal floods, ranked the world's most destructive floods from 1900 of the twentieth century beginning:

- 1900: Flood in Texas, USA, on September 8, 1900, due to a fourth grade hurricane (wind 233 km/h) that swept through Southeast Texas causing severe coastal flash floods, with violent waves hitting the city. As a result, more than 3600 houses were destroyed and 12,000 people died. This is the worst natural disaster in American history.
- 1931: A series of devastating floods in Central China in July 1931 are recorded as one of the deadliest natural disasters in modern history. Heavy rain combined with tornadoes created flash floods in many catchment points and submerged the downstream basin of the Yangtze and Hoai Ha rivers. Fifty-two million people were affected and the terrible death was uncountable (500,000–1 million people), followed by famine and epidemic spreading.
- 1953: The coastal flash flood in the North Sea, Europe caused by the worst storm in history there on January 31, 1953, drowned 162,000 ha of the Netherlands, killing at least 1800 people, spread to the west causing the dam on the England east coast to break, flooding 65,000 ha of land, and 133 people died when the MV Princess Victoria ferry sank.
- 1999: The tragedy of Vargas, Venezuela, on December 14, 1999, occurred when torrential rains caused flash floods in this coastal lowland, causing deadly landslides that destroyed the Caraballeda town. The death toll is estimated at between 10,000 and 30,000, and 100 km of shore were completely destroyed in just a few days. It is the worst natural disaster of modern Venezuela.

We can also review some recent typical flash floods such as 2004 tropical storm Winnie that caused big rains that led to flash floods and mudslides that killed more than 200 people in the General Nakar town of Quezon province in East Manila, Philippines. A flash flood occurred on October 2005 in Fujian province, China, and caused 80 deaths and 39 injuries (Fig. 2.3).

Fig. 2.3 European city flash flood

In June 2006, a flash flood in the Northeastern US killed 12 people. In August 2006, a flash flood in Rajasthan state, India, caused 300 deaths. After that, stormy rains lasted for 2 days on May 23 and 24, 2007, causing flash floods and landslides in Southwestern China, killing 21 people and 11 others were reported missing. The flash flood and landslides that hit Ganzi and Liangshan districts in Sichuan province on May 24, 2007, killed 11 people and injured 5 others.

That same year, in June, great rains lasted for a few days causing a flash flood in Andhra Pradesh state of South India, killing at least 45 people and forcing tens of thousands to leave their homes for evacuation. In October 2009, big rains for 4 days also caused flash floods in Karnataka and Andhra Pradesh, South India, killing more than 172 people.

Since the 1970s and 1980s, the problem of flash floods has attracted a lot of attention from scientists and leaders in the world. The World Meteorological Organization (WMO) has enhanced the establishment of flash flood warning systems through a series of projects and has initially seen significant successes such as the ARLERT system used in the USA and some other countries.

2.3.2   Vietnam Flash Floods

Recently, climate change, extreme weather developments, and natural disasters in the world and in Vietnam have become increasingly severe and complicated. Located in the tropical monsoon climate, Vietnam receives many natural advantages but also suffers from bad weather impacts such as flash floods.

The provinces in North Vietnam such as Ha Giang, Tuyen Quang, Dien Bien, Lai Chau, and Son La, the Central Highland provinces such as Dak Lak, Dak Nong, and Lam Dong, the Southeastern provinces such as Binh Phuoc and Tay Ninh, and the North Central mountainous provinces such as Thanh Hoa, Nghe An, Ha Tinh, Quang Binh, Quang Nam, and Quang Ngai are the places at high risk of flash flood due to the natural environmental conditions and human impacts.

Flash flooding is a type of natural disaster that has been increasingly occurring in the Vietnam mountainous areas. In addition the great the economic and human damages caused by flash floods, the damage to the living environment is huge. Flash floods and landslides have occurred in most of the Vietnam mountainous and hilly provinces, and many flash floods and large mudflows have caused great damage to the economy, society, and environment of the whole region.

According to the statistical summary, from 1953 to 2016, there were 448 flash floods and landslides (7 times/year on average). Particularly from 2000 to 2019, there were more than 320 flash floods and landslides affecting residential areas, causing more than 1000 casualties, increasing the average rate to 12–16 times/year. It is obvious that the frequency, intensity, and damage level of flash floods are increasing at an alarming rate. Fortunately, the loss of human life is not great because the majority of flash floods and landslides occur in far off, sparsely populated mountainous areas (Fig. 2.4).

Vietnam has recorded historic flash floods in the Northern mountainous provinces, North Central and Central Highlands, such as the flash flood in Quan Can fountain (Thai Nguyen), October 20, 1969, where 26 people died and many were

Fig. 2.4 Flash flood at Lai Châu province, Vietnam

injured; a flood in Lai Chau town, on June 27, 1990, swept away 300 people, 104 dead, and 200 injured, damaged 14,300 m² of houses, and 300 ha of rice fields were destroyed; a flash flood in Son La town, on July 27, 1991, killed 21 people, 11 missing people, 100 houses were swept away, 762 houses were flooded, 5000 ha of rice, and hundreds of hectares of crops were damaged.

Flash flood in Dinh river basin of Ham Tan area, (Binh Thuan province), July 1999, sank 80 boats, killed 27 people, flooded 11,101 houses, seriously damaged; Flood on July 15, 2000 in Sa Pa (Lao Cai province) killed 20 people and injured 25; Flood on October 3, 2000 in Nâm Cuông village in Sin Ho district (Lai Chau province) killed 39 people and injured 18 people; The flash flood on August 16, 2002 in Bac Quang and Xin Man districts (Ha Giang province), killed 21 people and injured 8;

The biggest historic flash flood in Ha Tinh made over 50% to 80% of communes in Huong Son, Huong Khe, Vu Quang of the province flooded 3–4 m deep, killing 83 people and missing, and 117 injured; A historic flash flood in 2004 on 2 communes Du Tien, Du Gia of Yen Minh district, Ha Giang province and Bao Lam district, Cao Bang province, killed 56 people; Landslides in Lao Cai left 48 people dead and missing and 16 injured, of which the whole family died; Flash flood in Nghe An on 12/8/2005 killed 16 people…

Extreme weather in 2008, reflected in the number of storms No.4 and No.6, caused massive flash floods and landslides in many parts of the Northern mountainous provinces such as Lao Cai, Yen Bai, Son La, Lang Son. Ha Giang, Cao Bang, Quang Ninh, Bac Giang killed 246 people and were missing, more than 200 injured, property damage was estimated at more than VND 3229 billion.

In particular, only in 2018, with the anomalous pattern of climate change, the Vietnam Northern and North Central regions had 14 flash floods, landslides, causing 82 dead and miss, accounting for 70% of national disaster damage with losses amounting to over VND 10,000 billion. In this year 2020, only in the Northwest regions, severe flash floods and landslides from June 2020 in Muong La and Son La cause 23 deaths; Mu Cang Chai, Tram Tau, Yen Bai killed 53, more than 1500 houses were damaged and transportation infrastructure was widely damaged.

Especially dramatic is that in 3 months September October and November of 2020 in the mountainous and coastal provinces of Central Vietnam (from Thanh Hoa to Phu Yen), after 6 consecutive storms [Saudel (No.8), Molave (No.9), Goni (No.10), Khanun (No.11), Atsani (No.12), Vamco (No.13) – Vietnam names storms by ordinal number in a year], 242 people died and were missing, 222 people were injured, over 200,000 houses were damaged, more than 3000 cattle and 600,000 poultry died, 45,000 ha of rice and 22.3 thousand ha of plant crops were lost, many infrastructure projects were seriously damaged, with a total economic loss due to storms and floods in recent years are nearly 28,800 billion VND.

This "cataclysm" with "floods overlap floods" in October and November 2020 had damaged areas spread across 10 Central provinces and the Central Highlands, caused big sabotage on 12 national routes, more than 17,400 m of local roads eroded, of which the most serious are Thua Thien Hue, Quang Tri and Quang Nam provinces.

Storms and floods ragingly damage Vietnam mountainous and coastal provinces that are the newest evidence of a worrying trend that the natural risks, already very dangerous, are becoming increasingly aggravated by high-speed urbanization, economic development, and climate change. There have been many alarming statistics on the vulnerability of this region and those most affected, with an estimated 12 million people at risk of heavy storms and floods and more than 35% of habitats is currently situated in high-risk areas for flash floods and landslides, with an average of $852 million vnđ per year – equivalent to 0.5% of GDP, and 316,000 jobs in key economic sectors are affected.

Not only the problems of economic – social – environmental management, disaster prevention, but this kind of flood is posing the serious threats to energy security, food security, and education. The gloomy picture of this kind of natural calamity with somewhat human factors has set an alarm off to act immediately regarding flash floods prevention in the overall context of natural disaster response and climate change to avoid further loss of people and assets and prevent floods from growing bigger and fiercer before it is too late.

2.4   Cause of Flash Flood

Flash floods are impacted by combination types, natural conditions, and forms of human activities in the basin. In essence, the factors can be classified into three groups depending on their fluctuation rate (Fig. 2.5).

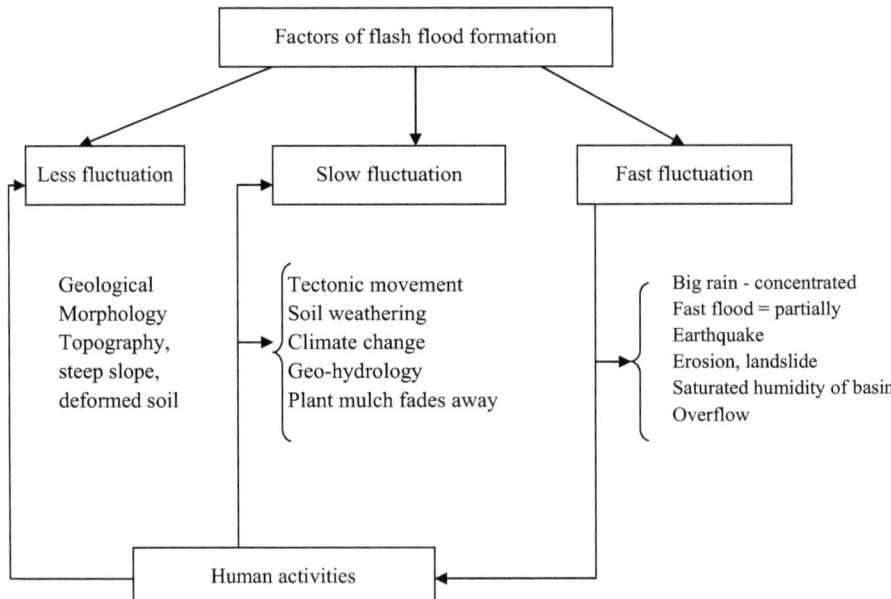

Fig. 2.5 Factors of flash flood formation

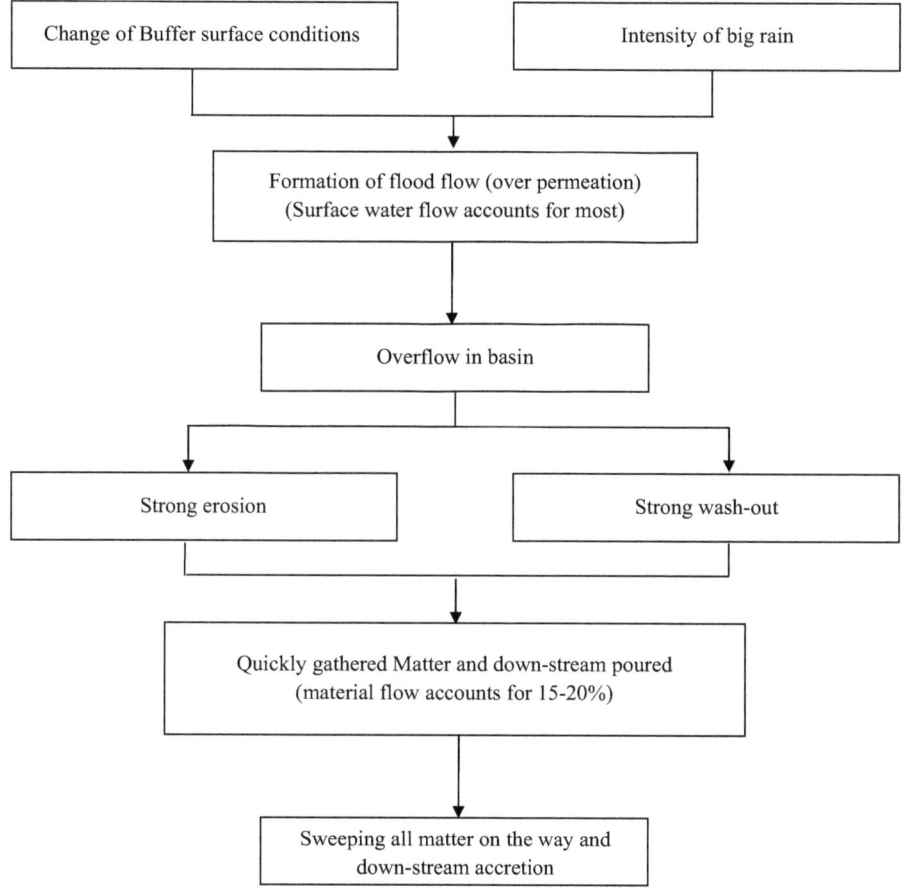

Fig. 2.6 The mechanism of flash flood formation

The patterns of human activities in the basin can affect all three factor groups: little change, slow and fast fluctuation. But the most obvious fluctuation is the factor group that change rapidly. This is a group of indicators often chosen as the characteristics to distinguish flash flood from normal flood (Fig. 2.6).

Brief, Flash flood occurred due to many reasons, which can be attributed to three main groups of causes, namely (1) Due to unfavorable natural and objective conditions; (2) Due to the shortcomings of the works and forecasting systems, prevention mechanisms in flood prone areas; (3) Due to human activities.

2.4.1 Natural Cause of Flash Flood Formation

As recognized, some Vietnam mountainous provinces on Northwest, North & Middle Central, Central Highlands and the Southeastern and other worldwide regions with the mentioned similar typical feature are at high risk of flash floods due to the conditions, natural topography, and human environmental impacts.

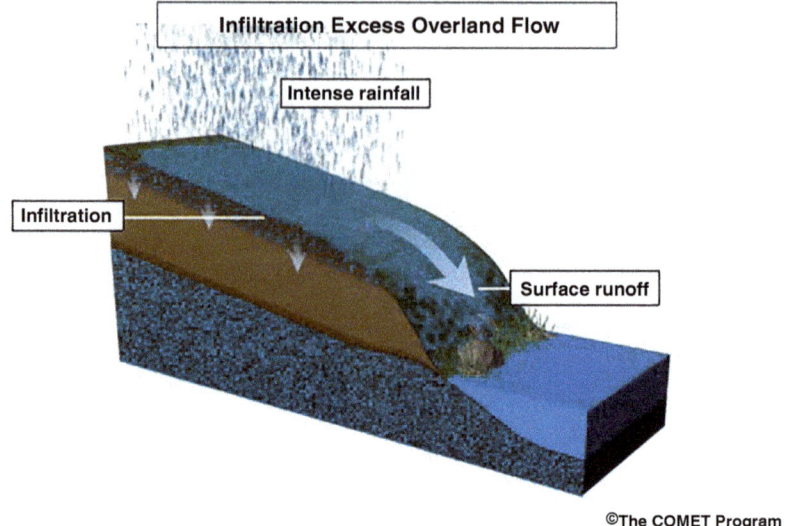

©The COMET Program

Fig. 2.7 Flow infiltration on land

Flash floods can occur under certain conditions and causes. Most commonly flash floods occur when it rains rapidly on saturated soil or dry soils with poor absorption. The water flow concentrates by accumulating the surrounding currents that overflow into the basin and then link together to create a concentrated flow with a larger volume, often fast flowing with debris washed away in surface water.

Flash floods most often occur in dry areas that have previously received significant rainfall, but can also occur anywhere downstream from upstream rainfall causing water overflows. Flash floods also occur after volcanic eruptions, from melted glaciers, and heavy rainfall due to storms and tropical depressions (Fig. 2.7).

2.4.2 Causes of the Short-Comings and Inadequacies

– **Of the works**

Infrastructure systems such as electricity and water works, roads, and stations, particularly in Vietnam and many other countries, are spread everywhere to serve the needs of production, management, and people's live. However, the works for treatment, prevention, and protection such as dykes, regulating culverts, drainage canals, anti-landslide embankments, overflow retaining walls, monitoring landmarks, and flood avoidance houses, have almost no investment interest, they are also fragmented, lack uniformity, often forgotten and not taken care of; therefore, they cannot be brought into full execution for actual effectiveness for flood prevention.

Even during flash floods, these infrastructure and public facilities are in danger of being swept away, meaning the service delivery can be disrupted during times of need. Severe flooding is directly affecting the authority offices, hospitals, clinics, and schools in the area, so there are also no emergency places to shelter, provide relief or medical intervention. The clean water supply network is also not designed to ensure survival during and after the flood. The epidemic prevention does not have on-site staff, and therefore must wait for remote mobilization.

More than 33% of the electric transmission networks of Vietnam and other countries are located in the forest–mountains areas, and thus are at risk of damage from trees falling due to storms. Transformer stations and transmission lines for electricity located in flood plains, when flooded, destroyed, collapsed, broken or cut off roads, are almost paralyzed and unable to provide the necessary energy. Actions for the repair, communications, lighting, machining, etc., are often almost completely cut off, isolating the vulnerable areas.

It can be said that the infrastructure network for monitoring, preventing, forecasting, warning, evacuating, sheltering, logistics, and recovering has not been well organized to ensure living appropriately with floods.

– **Of the warning system**

The current risk management solutions are not enough, and although the risk management programs of Vietnam as well as many other developing countries have made great progress over the past decade, significant challenges remain. One of the shortcomings is inefficient enforcement of construction standards, safety regulations, and maintenance of infrastructure systems.

Different from weather forecast information or remote disaster warnings which are still relatively effective, for flash floods, without early forecasts and timely warnings, they will be considered as nothing. In fact, in many places – not only in Vietnam but also in many other countries with the same economic, social, – scientific, and technological circumstances – this problem remains a difficult challenge to solve.

Our flood forecast is still based on interpretation/analysis of geostationary satellite images and monitored data of meteo-hydrological stations combined with analysis of flood alarming models according to algorithms and experiential coefficients in comparison with international and regional forecast information. Such a long chain process never catches up with the unpredictable, lightning-fast developments of flash floods.

In addition to Vietnam, many countries of the world still think timely and accurate forecasting of flash floods is very difficult under the current technical conditions plus the increasingly complex and rapid developments of flash floods in accordance with the momentum of climate change.

– **Of the reserved conditions**

The localities at high flash flood risk, despite having experience in coping and encountering flash floods, still do not have a sufficient system for monitoring, warning, prevention, safety, relocation, gathering, evacuation, rescue, relief, and rehabilitation together with structural solutions (watershed forest, infrastructure works, intervention system) and non-structural solutions (education, training, environmental awareness, media, communication) to respond to this disaster type.

Vietnam has a motto "4 on-site" (Command/Force/Materials/Logistics), and although very flexible and correct, it is not seriously implemented by many localities due to insufficient resources or lack of awareness of this policy's importance. Everything is habitually looked for from superior power – more regular, more powerful and no worry about the duty of higher authority, so long as the usual scenario of flash floods keep "coming for appointment" again, then when the violent flood arrives, everyone goes to District or to Province to find shelter, receive relief, and wait until the water recedes, and continue to comeback there for life restoration and production, thence continue to wait for the next flood.

2.4.3 Cause from Human Impacts

In this article, damages of flash flood caused by human impacts are the most concerning, because any artificial work, especially in mountainous areas, river upstream, all affect the environment, including the effects on the chemical–physical properties of geological layers in a concrete area.

Roadwork and construction of factories, farms, lakes, dams, mining, and civil works all need to be renovated, changed to a certain area, which is forests or hills and mountains. The deforestation for work surfaces, corridors, protective belts, and project roads has all erased the buffer surface and natural forest trees in the area, increasing the risk of water spillage without stopping or having a braking force for flash flood flows. Both legal and illegal watershed deforestations destroy the water retention capacity of natural land resources, which are the leading culprits of flash floods.

It is possible that flash floods in particular and floods in general are the premise factor and driving force for landslides, and these two phenomenon are closely linked, and the greatest impact of watershed natural forests is to greatly limit that casualty. The area of primary forest is narrowed, replaced by planted forests in those positions, thus changing the water-holding capacity, and the slope of the hillsides and mountains is cut high, losing the stable equilibrium of the slope created a million years ago.

The destruction of natural forests and vegetation cover for the work surface engineering of the mineral mines and the hydro-power plant & dam-lake are the two biggest flood-damaging factors, which have made the land in these locations more barren and deserted, and a new plantation will not bring any watershed protection effect. The floodwaters that gathered on the corridors of the basins were not restrained by the protection forests and had to force themselves to concentrate increasingly aggressive forces to flow to the lower areas for destruction.

Planted forests can never be effective in replacing primary forests and natural vegetation with big root-sets, wide spreads, large connections, and a uniform network. In addition, the planted forests are easy to uproot, creating a direct impact to "activate" the mechanism of massive soil–rock movement, leading to landslides accompanying flash flood, and when drifting, increases the volume of solids and kinetic energy for the flood.

Humorously, in many places, many people still think that trees of coffee, macadamia, rubber, melaleuca, acacia, etc., are the greening plants for barren hills that will prevent natural disasters, with dual effects of economy and environment. They do not understand that only natural forests, whether primary or secondary, with many high and low tree canopies and interwoven vegetation by closely dense roots on the forest floor will have the effect of real protection–prevention, and that "reforesting" and "greening the area" are misused words for sophistry and offer no actual land protection or flood and natural disaster prevention.

Hydro-power dam, theoretically, in addition to dynamo rotation of generator sets and returning output water down-stream, can also work to regulate water and prevent floods when properly managed and operated. However, that is only in principle, and with the unpredictable extreme weather characteristics of the climate change era, it is a double-edged knife.

When there are observable, predictable fluctuations of the normal weather, the dam spillway and flood outlet mechanism of the hydro-electric dam will operate effectively as the designed scenario, and the reservoir actively regulating flood water will moderate the flow rate, intensity, and flood flow kinetic energy to downstream.

Woefully, in extreme weather conditions with unpredictable storms and irregular rains, there is no way the lakes and dams can solve floods that exceed alarming level many times and chaotically have to flood continuously with their maximum rate and what is calculated on the design is often not fully capable of response to the spillway system, the discharge outlet in correspondence with the level of flash anomalies of upstream and on-site water accumulations for flash floods.

On the European calculation basis of hydro-electric dams, an assessment on capacity and physical water treatment ability of the dam found that all dams (both hydro-power, irrigation, and water supply) in fact partly helped the flood mitigation in all European countries. However, the extent depends on the reservoir volume and the operation process of the lake. The large, well-managed forecasted lakes that can release water before flooding will greatly reduce downstream floods, while small lakes have no or little capacity to reduce flooding.

Of course, those facts only happen under normal conditions and even if there is an anomaly, they happen according to the regular rules. In addition, Europe does not experience many hidden hazardous events from monsoons, tropical convergence, sub-equatorial storms, and unstable mountainous terrain like in Vietnam or in many other tropical and sub-tropical countries.

In Vietnam, for example, Hoa Binh hydro-electricity on Da River or Da Nhim, Tri An on Da Dung river, Dong Nai or Dau Tieng lake of Saigon river, the hydro-power dams or large irrigation, hydrologic treatment structures with modern, major scale are well-managed and have real and strict regulating, but the rest of minor lakes, dams, and hydro-electricity has less informed, irresponsible management and regulation, which only causes further harm to the environment. In the current world trend, hydro-electricity is considered as a renewable energy source but soon became "dirty" due to its negative impacts on the environment, and is thus not a form of clean energy anymore.

At present, many countries in North America and Europe are in aggressive campaigns to demolish the hidden-damaged dams, while sadly, in Vietnam, this "business disease" has not been much reduced. Although the State policy is not in favor of small hydro power of less than 15 MW, but unfortunately localities seem excited about investment projects that bear this potential catastrophe and the whole country has been, is, and will be building more than 800 hydro-electric plants like so.

The great impact of hydro-electricity is already worrying, but this kind of disaster will multiply many times when there are small hydro-power dams scattered along the lower steps from top to bottom of the stream-bed ladder to "every-house, every-people" to profit by any space of the river bed for energy production and legal deforestation business (some places are only 15 km of river distance, but up to 4 factories hustle on 4 steps of the conductor-bed elevation).

Moreover, those plants always prefer to integrate maximum water in reservoir to ensure "natural capital" for power generation, making the flood discharge scenario increase the volume suddenly, accelerating, resonating potential and kinetic energy that sets up a flash flood chain like a domino effect from the highest dam lake dumping water that rushes down with progressive speed and volume, and the last lake dam will eventually "burst" and let rushing water discharge, with the extremely large possibility of dam failure.

The 4-steps ladder of Rao Trang hydro-power (A Lin 1, A Lin 2, Rao Trang 3 and 4) on the Rao Trang river (other name: Khe Bun) located in the Vietnam Central Middle ecological core is a clear example. The unreasonable problem here is that it is better to have no dams so that the savage flow will overflow or flood naturally to wild downstream over a large amount of time rather than filling up with water to coincide with the large floods then exceedingly "spend" all at once. Therefore, deforestation for wooden business or hydro-electricity and the unsustainable development in the manner of massively pursuing GDP growth while serially devastating nature for economic, industrial, and energy projects in the regions at flash flood risk are the culprits causing this man-made disaster (Fig. 2.8).

As a survey report by World Bank, if the trend of rapid economic development continues in high-risk areas like today, the damage caused by natural disasters will increase. It is time to take a new approach to balance risks and opportunities so that every country's key regions can continue to motivate growth, while also ensuring

Fig. 2.8 Forests cleared up in the traffic corridors and installed hydro-electricity works

resilience to shocks. The countries need to be sustainably developed into a space worth living, but the environment cannot be exchanged for economic growth.

2.5   Why do Flash Floods Cause Great Catastrophes?

The reasons that makes areas unable to withstand flash floods are mainly from the above cause groups, leading to their formidable destructive power. These aforesaid causes are both the premise condition and the motivation to form flash floods, and at the same time, they also explains why flash floods cause such tragic catastrophes. According to the relative calculation, flash floods are at the top of the list of natural disasters that cause the loss of human life and environment damage, as the largest, fastest, and most concentrated, mainly because their irregularity, aggression, super quickness, crisis, destruction, cruelty, and extremely unpredictability make it difficult to escape.

In addition to the three mentioned causes, other reasons are natural anomalies, the objective conditions of insufficient investment resources, and human destruction of the environment leads to the destructive power of floods. That is the basic part, but the end part of this causality correlation is the result of flash flood; however, there are other reasons to explain why flash floods can cause such tragic disasters over the human resilience. Any type of natural disaster, including flash floods, any event no matter how terrible, and people still find a way to live with it by any means. But why do people still suffer from disasters because there are not enough effective measures to cope?

With enough painful experience, enough science and technology, enough knowledge to monitor, manage, forecast, enough spiritual–material resources, perhaps the only lack is the scientific and reasonable organization, synchronous coordination, rational use of resources in hand to proactively respond effectively. These are the purely subjective reasons for flash floods to turn into such crazy, nasty natural disasters:

- The first is the rapid and chaotic development, lack of organization and scientific inadequacy of infrastructure in vulnerable areas, especially in mountainous and coastal areas. These infrastructure projects lack coordination, supervision and are not well-organized, and they lack synchronization, interaction, do not consider the predicted extreme impact, are fragmented, spontaneous, unilateral, on demand without vision, and most important, have a shortage of risk assessment and environment impact. So every material can be destroyed in natural disasters.
- Second is the policy of organizing the weather observation, forecasting, disaster management, which has a lack of logical and scientific cohesion, lack of real-time data interaction, and lack of serious implementation from central to local grade. This needs to be improved in terms of expertise, technology, and equipment, especially the mechanisms of management and enforcement in order to have a policy, legal framework, and a competent disaster management board.
- Third, the lack of coordination among management units leads to the situation that each manager issues their own policies and regulations, and the method sets for collecting, analyzing, and handling emergency situations are reciprocally different in natural disaster issues, causing overlapping, confusing the implementers at all grades who do not know how to apply in practice, and are often "improvised as per changes" rather than following a unified process.

If we do not improve these three issues, the responding forces will be more confused and disturbed when they become available and the damage caused by natural disasters in the coming years will be even more serious.

2.6   The Main Stages of Flash Flood Formation

Research results show that the formation of flash floods goes through the following stages:

- *Stage 1:* Heavy rains with high intensity, cause high-surface floods and especially overflow on small basins in steep mountainous areas with little forest cover, much exploitation, potentially favorable conditions for erosion, washed away rock, mud, sand, trees, but the conductor-bed in poor drainage.
- *Stage 2:* The flood causes erosion, washout, landslide, slippage, strong landslide in the basin surface, swept away solid matter, then flood flow changes basically in quality, becoming liquid–solid (water, mud, rocks, trees, etc.) concentrated in

the main stream. The flood then has a total weight greater than that generated by the floodwater alone.

- *Stage 3:* The flood-forming area is the upstream part of the river basin with a steep slope, usually accounting for 2/3 of the basin area. Here, the main processes of surface runoff, erosion, and ground washout occur most strongly. The process of flood flow concentration also happens at the same time, but not strongly.
- *Stage 4:* The area of concentrated flash floods, where deep erosion still occurs strongly, causes the slippages, landslides, sweeps away trees, in temporary congestion and then massively breaks.
- *Stage 5:* Flood objective area: the place most strongly swept is at the end of the slope when potential energy has turned into kinetic energy, in which deep erosion, slippage, and landslide still occur at high intensity on the entry head of the valley. Phenomenon of floods and sediments occurs most strongly at the end of the valley before flash floods can escape the main stream.

This stage division is aimed to easily analyze and evaluate, but it is relative. These stages usually happen quite quickly to very quickly; sometimes the previous stage has not ended, and the next one suddenly comes.

2.7 Basic Feature of Flash Flood

(a) **Surprise**

The time since the appearance of an increased water level of a current in stream and small river overflow and gather to the flood peak is very short, usually within a few hours. Therefore, it is often difficult to effectively predict and warn about flash floods at the current technical and professional level.

(b) **Fastness and fierceness**

Flash floods usually come and exist in a very short time; usually end after 10–18 h, and rarely last more than 1 day. Large floodwaters erode and wash away huge amounts of solid matter from steep slopes and become a mixed stream of mud–water–solid substances that gather almost simultaneously and move together, thus having a very fast speed. The total mass is large and the flood peak is very high – different from normal flood in equivalent rainy conditions (sometimes two to five times higher) due to the different mechanism of formation and movement.

Flash floods are brutal because they come quickly then stop quickly. Thus, in order to reduce or eliminate the short-term nature of flash floods, measures must address the predominant dynamics during rising floods – which are primarily toward the concentration period of flood flows in the upstream basin, from there, it also reduces the severity of the flood in the upstream catchment (reducing the flood peak, flood frequency up and down, the flow rate of rivers and streams).

(c) **Condensedness**

Flash flood flows are different from normal flood currents because of a very large proportion of solid substances. During the process of formation and movement, the proportion of solid objects in flash flood currents successively increases – most strongly in zone 2 – when moving from high mountains (the stage over slopes) down the valley.

The solids amount usually accounts for 3–10% of flow volume, even more than 10% in the flood current like muddy and rocky flood. Such a flow, in terms of its forming nature and dynamics, is different in quality from the normal flood. Flash flood flow is the intermediate phase between liquid and solid matter. This feature and impact is due to the basin surface being eroded and having slid, increasing the amount of solid matter in flood and the process of sliding motion.

(d) **"Domino" chain motion**

One of the basic properties of flash flood formation and development is its stylist domino chain reaction. That is, during the course of any flood, there will be a serial collapse of the chain when the flood flow deforms and impacts on the materials in its path. Therefore, as its severity increases, its flashiness increases. The flow kinetic energy is maximized until the potential energy is completely annihilated in a lagoon or valley or large river where flood flow enters and stops.

(e) **Acceleration**

The initial speed of the flood is not very high, but the acceleration increased rapidly as per time. For example, in a small catchment, flood flows can peak extremely quickly, while at a cross section of the channel current, the flow between them can reach nearly 20 m^3/s.

2.8 Classification of Flash Floods

Classification of flash floods is based on rule uniformity of formation and development. Scientists have studied and classified five flash floods as follows:

1. **Congestive flow flash floods**

This type of flood occurs with relatively high intensity and speed, has a flood amplitude with large flood depth, and carries a lot of different materials (rubbish, mud, sand, and plants). This type of flood is formed in river valleys, water-catchment areas between mountains or fields, etc., and floodwaters are blocked due to many causes, for example, flash floods that occurred in Dien Bien Phu City (1996), Son La Town (1989), on Nam Cuong Stream (Bac Can, 1981), Lang Son Town (1986), Huong Khe, Huong Son (Ha Tinh, 2002, 2007), and many flash floods have occurred recently in the Southeast and the Central Highlands.

This stylist flash flood usually occurs in mountainous, hilly areas where the basins that accumulate floods are often covered by plants with shallow roots or unstable structural works or flood flow through unsustainable geological beds, in areas where even the houses, buildings, and vehicles of the city, towns are low-lying and coastal. This stylist flood is relatively common on types of terrain with many obstacles on flood flow and causes the representative damage of flash floods.

2. Slope side flash floods

This is a type of flood with a very high flow rate, fast up and down, bringing a lot of material from the hillside and basin side. Slope flash floods occur mainly on steep slopes in areas where surface water is concentrated. Slope flash floods have occurred in Quang Ninh, Hoang Lien Son, and North Central. In mountainous and coastal areas, this type of flood is frequent.

3. Rock–mud–timber–plant flash floods

It is a particular type of flash flood by water currents with a large amount of sediment and high kinetic energy. Mud–rock floods arise from the upstream small fountains, mostly tributaries of grades I and II, where the soil and rock are strongly eroded and poured out into the estuaries. For example, large mud–rock floods occurred in Muong Lay Town (Lai Chau, 1996) and Du Tien (Ha Giang, 2004). According to technical and traditional classification, only when density of rock and soil in the water flow is greater than 60% of mass, thence is called mud–rock flood.

4. Flash floods caused by breakage of dams, dikes, and reservoirs

This flash flood is caused by an immediate discharge or rupture of a lake, dam, dyke, or a hydro-electric or irrigational construction. Flash floods of this type are very destructive in large areas on downstream because their poured floodwaters on rivers and streams with rugged terrain will create tsunamis sometimes tens of meters high, drowning everything.

Small hydro-electric dams emerged as the biggest factor causing breakage of lake dams or sudden water discharge to save the reservoir, releasing huge water volumes in a short time making flash floods; coaxial tubing flood for downstream. The dam break is so horrible that it has led many developed countries to carry out big campaigns to protect the ecological environment and life safety. On the Vietnam Energy Forum in 2016, Engineering MS Nguyen Huynh Thuat does not support the construction of hydroelectric dams because of their harms to the environment, and he said "Among the casualties of human structures, dams are the deadliest."

5. Complex flash floods

The type of flood with high flow velocity, great flood intensity, and relatively large depth of inundation. Complex flash floods are mediated by congested flash floods and side flash floods. These include the mixed flash floods in Quan Cay (Thai Nguyen, 1969), Nam Cuoi (2000), and Truong Son commune (Quang Binh, 1993). This type of flash flood occurs commonly in mountainous areas and often causes large losses in life and property. One basic feature is that mixed flash floods occur in small and medium sized mixed water basins between mountains and hills or an accumulating shelf located on a steep riverbed. Note that the term mixed flash flood here differs from the mixed type which many researchers consider to be the sum of flash floods and mudflat floods.

These are five types of flash floods classified according to the characteristics of the flash flood formation and flow material to satisfy the technical criteria, academic content, and professional solutions. But more generally, in reports, plans, and especially in establishing flash flood risk maps, people only rely on flood movement and development mechanism to classify into three types of flash floods, which are (1) congestive flow flash flood, (2) slope side flash flood, and (3) complex flash flood.

2.9   The Basic Criteria to Identify Flash Floods

For the design and implementation of any type of structural measure, even for non-structural measures, the basic characteristics of flash floods are the most important basis, in addition to knowledge of the formation site, movement, flood-prone areas, and flood features.

The basic criteria to define flash floods are:

– Time of occurrence, of floods up and down and the whole flood, scene process.
– Flood peak, flood amplitude, average and maximum flow velocity.
– Average and maximum flood intensity.
– Total volume, material composition in flood (liquid, solid), physical and mechanical characteristics of the stream.
– Flood concentration time, flood transference time, transmission capacity of flood flow.
– Composition of solids, particle matter, and the particle distribution in the flood stream.
– The current momentum and impact when encountering an obstacle.
– The geometric dimension of the conductor-bed (overflowing or tubing).
– Hydro-dynamic pressure during dam failure (irrigation dams or dams newly formed due to flow movements), and temporarily stagnant during flash floods.
– The inertial speed when the flood increases and stops, depending on the flash flood structure.

2.10   Environment Impact Assessment of Flash Floods

The environment affected by flash floods includes natural and socio-humanist.

2.10.1   Natural Environment

- Environment of land resources.
- Environment of forest resources: Areas of high importance to biodiversity, habitat of many threatened species (tropical forests).
- Ecosystem environment of soil, forest, and water.
- Environment of small river basin, streams and characteristics of hydrological regime (main river and tributaries, flow volume in flood and dry season), surface water areas (lakes and lagoons), self-cleaning ability of water.
- Quality of water sources (surface water, under-ground water).

2.10.2   Socio-Humanist Environment

- Residential points inside and near areas of flash and tubing floods.
- Current economic industries in the region have to change their structure or activity level (agriculture, forestry, etc.) because of flash floods.
- Current land possession system.
- The system of life ensuring and other income sources of local people, the distribution of local sources of income.
- Environmental sanitation and community health.

2.10.3   Analyze and Forecast the Environmental Impacts

2.10.3.1   Impact on Natural Factors

(a) **Impact on soil quality**

- Flash floods cause erosion, washing-out, slippage, strong landslides in the basin surface, and sweeping away of the solid matter such as soil, sand, pebbles, and even big rocks to the hill foot.
- Fertile soil layers are washed-away and soil resources are degraded by erosion and ground abrasion.
- The areas below the hillsides and low-lying areas are swept up by soil & rocks, garden and fields are covered with soil & rock, no longer cultivated.
- The structural works were accumulated by rivers, reservoirs, the depth was shallow, and the lake bottom was deformed.

(b) **Impact on water quality**

- Flash floods fundamentally change the flow nature into a solid + liquid one (water, mud, rocks, trees, etc.), which flow into fountains then into main rivers. Thence the flood is greater than its total volume first generated.

– Pollute the surface water sources: flash floods drag soil, rocks from steep slopes plus pesticides and fertilizers used in agriculture into rivers and fountains that pollute water sources and increase turbidity, adding solids, contaminant contents, chemicals, and pesticides to water sources. An increase in the amount of fertilizers and pesticides will increase the possibility of pollution discharged into the soil and water environment, and when the receiving environment is polluted and unable to self-clean, then more severe pollution will occur to the environment.

– Pollute the groundwater: in areas where during passing flash floods, people's underground wells are flooded with pollutants, contaminating the water and making them unusable.

(c) **Impact on the ecosystem**

• *Terrestrial ecosystems*

– Plants in the damaged zone are washed away, broken, causing zonal erosion, slide, become deserted, changing the landscape of a large area.
– Vegetation in the damaged zone will be destroyed after flood, leading to the flora deterioration and the destruction of animals living in this environment. Some species of flora and fauna in this ecosystem will be lost or decreased.

• *Aquatic ecosystems:*

• Aquatic systems are also affected by flooding because of broken plants, trees, and mud/rocks entering rivers and fountains.

– The fallen rock and eroded soil increase the turbidity, reducing dissolved oxygen in water or water surface, so some species of flora and fauna living in the area will be reduced or lost.
– The decreasing vegetation cover affects the water environment in the vicinity, the rate of water evaporation increases rapidly, leading to drought and degradation of the aquatic ecosystem.
– Flash floods bring pollutants such as fertilizers and pesticides into the water; the aqua-creatures either die or are destroyed. The surviving species will absorb the toxins into the body, then forward them into the human body through the food chain.

2.10.3.2 Impact on Socio-Economy

(a) **Impact on human life**

In areas of flash flood occurrence, people's lives are often threatened, especially people living in the area at the foot of the mountains and the hills. Flash floods affect people's psychology thereafter: People may leave their residences to move to a new places that can be safer for their lives and properties, but this makes it difficult for a stable life.

Table 2.1 Evaluate the impact of flash/tubing floods

Impacted objects	Impact level			
	Very strong	Strong	Medium	Weak
Soil environment	X			
Ecosystem		X		
Water environment	X			
Traffic system	X			
Communication system			X	
Human's life and property	X			

(b) **Damage to Property**

Houses collapse, are possibly buried or swept away, and people have no place to live. Property is often lost after flash floods. Crops are destroyed, areas are swept away, rice fields and upland fields are destroyed, unable to reproduce. Because the animals were washed away and died, people will face many difficulties in life stabilization, leading to poverty.

Systems of road, bridges, and communication are destroyed. Landslides on traffic can occur sometimes on both positive and negative slopes, so many sections of the road are completely destroyed.

(c) **Increased epidemics**

When a flash flood occurs, a large part of the waste is washed away by the flood and flows into the surrounding water source. For residential areas located along rivers and fountains, due to the narrow space and poor drainage conditions, the level of water pollution increases, and the dirty water becomes a carrying route for microbes and pathogens to go everywhere.

Areas after flash floods can be sources of epidemics, extremely powerful microorganisms, decomposing leaves, and rotten animal carcasses, causing serious environment pollution.

The water source in many places is also contaminated, and if not treated, it can cause intestinal diseases such as cholera, dysentery, and typhoid In particular, the risk of cholera is very worrying because this disease often outbreaks in Asian countries (Table 2.1).

2.11   Signs for Flash Flood Prediction

It is necessary to observe the changes appearing around the living area such as sewages, drains on steep slopes (especially the watershed or catchment), landslide traces, collapsed trees, etc., or if the house window is stuck, unable to open, new cracks appear in the wall, ceiling, brick, or floor; any not-intact exterior wall, curb or staircase; expanded cracks appear on the ground or in the alley; broken ground water; the ground has phenomenon of blistery spots, precarious road; water splashes out from the ground in different places; fences, walls, electric poles, trees are tilted or moved.

Attention to any change in water flow. If the water turns from clear to turbid, this is also a sign of an imminent landslide. The sound of rocks falling gradually increasing, the ground beginning to move down steeply, soil layers receding, strange noises, such as the sound of a tree breaking or a rock collision are indications of an imminent landslide, and it is absolutely necessary to quickly move out of the danger area and notify local authorities and neighbors so that timely aid could be obtained.

Brief, it is important to pay attention to the signs of the flood risk listed below:

- Where you live, heavy rain forecast (> 200 mm), black clouds above, house or place near the stream, and the stream is small but the water level changes rapidly.
- Houses at the foot of the hill/slope or on the mountainside, around where trees were destroyed for planting, the land was plowed.
- Strange sounds occur.
- Stream water near your home or place with signs of turbidity, brown or black.
- Receive accurate information about a dam damage nearby.

Chapter 3
Collection & Investigation on Flash Flood Characteristics (Case Study: Dak Lak Province)

Abstract This chapter collects and investigates the natural and socio-economic characteristics arising from the impact of flash floods using the case study of Dak Lak province. The chapter analyzes the natural conditions, including geographical location, topography, soil, weather, climate, hydrological regime of rivers and streams in the area, and socio-economic conditions. The chapter also discusses the situation of economy, population, production and life, and the current status of land use.

The possibility of flash floods due to conditions of topography, structural works, arrangement of agro-forestry, scale of industrialization, population, farming organization, land, dams abuse, and deforestation are mentioned. The chapter also lists the flash floods in the past throughout Dak Lak province and assesses the levels of destruction and their effects. With these insights, a comprehensive framework strategy for flash flood control solutions will be proposed for Dak Lak province and elsewhere.

Keywords Lak lake · Buon Ma Thuot basalt · Per capita income · Indiscriminate land exploitation

3.1 Resources & Natural Condition

Dak Lak, meaning "Lak lake," with "Dak" meaning "water" or "lake," is a province with the fourth largest area in the Center of the Vietnam Central Highlands. Dak Lak is the ninth largest administrative unit in Vietnam in terms of population with more than 1.9 million people, the 22nd in Gross regional domestic product (GRDP), 41st in GRDP per capita, and 37th in GRDP growth rate (the GRDP growth rate is over 9%).

The capital of Dak Lak is Buôn Ma Thuôt city, 1410 km from Hanoi, 647 km from Da Nang city and 350 km from Ho Chi Minh City. In 2003, Dak Lak province split into two provinces: Dak Lak and Dak Nong. Dak Lak is considered as one of the cradles nurturing the Central Highlands "Gong" Cultural Space, recognized by UNESCO as the world-recognized oral and intangible masterpiece.

L. H. Ba et al., *Flash Floods in Vietnam*,
https://doi.org/10.1007/978-3-031-10532-6_3

Dak Lak forest has the largest area and reserve in the country with many rare and precious wood species, many specialty trees of both economic and scientific value, distributed in favorable conditions, so regenerating forests can be quite dense. Minerals with different reserves, including kaolin clay and brick clay. Moreover, there are many other minerals such as gold, phosphorus, peat, and gemstones in the province that do not have large reserves but are distributed in many parts of the province.

The system of rivers and streams in the province is quite plentiful, relatively evenly distributed, but due to the steep topography, water storage capacity is poor, and small streams have almost no water in the dry season. In addition to the rich river system, there are many natural lakes and artificial lakes such as Lak, Ea Kao, Buôn Triêt, and Ea Sô lakes.

3.1.1   Geographical Location

– The North border contiguous to Gia Lai province.
– The South border contiguous to Lam Dong province.
– The East border contiguous to Phu Yen and Khanh Hoa provinces.
– The West border contiguous to Cambodia Kingdom and Dak Nong province.

The province has a 70 km border with Cambodia, with an average altitude of 400–800 m above sea level, a total natural surface of 13,125 km^2, located in the central area of the Central Highlands. The North has important traffic such as National Route No.14, NR No.26, NR No.27 connecting the Central Highlands provinces with Central Coast, in which National Route No.14 along the border of the two countries is very convenient for economic development in border areas combined with defense and security protection (Fig. 3.1).

3.1.2   Topography

Most of the area of the province is located in the West Truong Son Mountains, with a lowering elevation direction from Southeast to Northwest. The terrain is diverse with hills and mountains alternating with plains and valleys, generally divided into the following types of terrain:

3.1.2.1   Mountainous Terrain (of Great Interest, Especially Geomorphology)

Is it possible to create a matter block that temporarily stops the flow or not?

Fig. 3.1 Administrative map of Dak Lak province

Dak Lak topography and geo-morphology is diverse with mountains alternating between plains and valleys, divided into many relatively different sub-regions, such as Ea Sup semi-plateau, M'Drak plateau, Chu Yang Sin highlands, Krông Pak – Lak low land, to create features of specific biological resources such as:

– Chu Yang Sin high mountain area is located in the Southeast of the province with an area approximately 1/4 of the natural area of the whole province, separated between the Buon Ma Thuot plateau and Lang Bian plateau (Lam Dong). The region has many mounts higher than 1500 m, and the highest, Chu Yang Sin peak, is 2445 m, with a sharp peak, vertically steep and rugged terrain. This is the largest aqua-biologic area, the headwaters of large rivers such as Krông Ana and Krông Knô and is a region with evergreen forest vegetation.

– The low and medium mountain region Chư Dơ Jiu is located in the Northwest of the province, separating the Ba river valley (Gia Lai) and the Buon Ma Thuot plateau, with the average altitude of 600–700 m, the peak of Chư Dơ Jiu is at 1103 m. The topography is abraded, eroded, and the flora includes regenerated trees, thin forests, and agricultural land.

3.1.2.2 Highland Terrain

Occupying most of the natural surface of the province, the terrain is flat, National Route No.14 is almost the drainage divide-top, high in the middle and lower on two sides, the terrain is lower from the Northeast to the Southwest. The whole province has two big plateaus:

– *Buon Ma Thuot plateau* is a large plateau running from North to South over 90 km, from East to West 70 km. Nearly 800 m high in the North, 400 m high in the South, gradually sloping to the West to 300 m high. This is a fairly flat terrain, with an average slope of 3–8°. Most of this plateau is fertile basalt red soil and likely fully exploited.
– *M'Drak Plateau (Khanh Duong Plateau):* located in the East of the province, adjacent to Khanh Hoa province, average altitude 400–500 m. This plateau terrain is rugged, with high mountains in the East and South, the central zone has a concave topography – high around and lower in the center. Granite soil occupies most of the area with evergreen forest vegetation in high mountains and low mountain grassland and gentle hills.

3.1.2.3 Ea Soup Semi-Plateau Terrain

The Ea Soup semi-plateau is a large area in the Western part of Dak Lak province, adjacent to the highlands. The surface here is abraded, the topography is quite flat, the hills are slightly wavy, the average height is 180 m. There are a few rising mountains such as Yok Dôn, Chư M'Lanh. Most of the land in the Ea Soup Plateau is gray soil, thin, and characterized by plants such as deciduous dipterocarp forest in the dry season.

3.1.2.4 The Topography of Krông Pach – Lăk Low-Land

Located in the Southeastern part of the province, between the Buon Ma Thuot plateau and Chư Yang Sin high mountains, the average altitude is 400–500 m. This is the valley of the Serepok river basin forming the low-lying areas along the Krông Pach and Krông Ana rivers with the Lăk–Krông Ana green field about 20,000 ha in size. This is a lowland that is often flooded in September and October every year.

Table 3.1 Surface of land types in Dak Lak province

No	Typical group of land	Surface (ha)	Ratio in total (%)
1	Alluvial soil(*Fulvisols*)	14,708	1.12
2	Gley soil(*Gleysols*)	29,350	2.24
3	Gray soil (*Acrisols*)	579,309	44.14
4	Red basalt soil(*Ferrasol*)	311,340	23.72
5	Black soil (*Luvisols*)	38,694	3
6	Other soil groups	338,372	25.78
Total		1,311,773	100

Source: Dak Lak Province Environmental Status Report (2016)

3.1.3   Land Resource

In the case of Dak Lak, land is one of the great natural resources endowed. The whole province has a natural land surface of 1,312,537 ha, in which mainly alluvial soil, gley soil, gray soil, red basalt soil, black soil, and some other soil groups such as soil group of eroded inert rocks (Leptosols), fractured soil group (Vertisols), newly modified soil groups (Cambisols), soil groups with tight clay, differentiated mechanized (Planols), dark brown soil group (Phaeozems), and brown soil group (Lixisols) (Table 3.1).

3.1.3.1   Alluvial Soil Group (*Fulvisols*)

Alluvial soils with a surface of 14,708 ha, accounting for 1.12% of the total natural area. The soil is formed by alluvium deposits of rivers and streams, distributed along the Krông Ana and Krông Nô rivers. Soil property depends on the weathering products of soil-forming precursor matters of upstream regions of each basin, time, conditions, and locations of sedimentation.

3.1.3.2   Gley Soil Group (*Gleysols*)

Gley soil group covers a surface of 29,350 ha, accounting for 2.24% of the total natural area, concentrated in the lowlands of Lăk, Krông Ana districts and scattered in the wet-lands in year entire.

3.1.3.3   Grey Soil Group (*Acrisols*)

Grey soil group also known as strongly acidic, low active soil, covers an area of 579,309 hectares, accounting for 44.14% of the total natural area, distributed mostly in all provincial districts, is the largest soil group in Dak Lak province and distributed in many terrain shapes, but mainly on sloping land.

3.1.3.4 Red Soil Group (*Ferrasol*)

The surface is 311,340 ha, accounting for 23.72% of the total natural area, the second largest after the gray soil group, distributed massively in the Buon Ma Thuot basalt blocks. This soil group has classified units: reddish-brown on basalt (Fk), yellow-brown on basalt (Fu), which is the B-layer soil group with the clearest accumulation of aluminum.

The Buon Ma Thuot basalt block flowing from North to South, from East to West. The North of the plateau (Ea H'Leo) has an altitude of 800 m, the South altitude 400 m, the West 300 m (Cư M'gar district area). The surface of the plateau is very flat.

3.1.3.5 Black Soil Group (*Luvisols*)

Surface of 38,694 ha, accounting for 3% of the total natural area, distributed around volcano craters, the edge of basalt blocks and basalt valleys.

3.1.3.6 Other Soil Groups

The total surface of other soil groups is 338,372 ha, accounting for 25.78% of the total natural area.

3.1.4 Climate Characteristics

Situated in a tropical monsoon climate affected by two opposite monsoon systems: equatorial monsoons and northern hemisphere trade winds. At the same time, it is influenced by factors of elevation and topography, so it forms a climatic type of highland tropical monsoon or "tropical plateau."

3.1.4.1 Temperature Regime

The outstanding feature of the temperature regime is that the monthly radiation balance is always positive; the total radiation energy in the regions averages 230–350 kcal/cm^2/year. The total number of sunny hours is 2200–2600 h per year, the rainy season has 6–7 h average of sunshine/day (180–200 h/month).

The annual average temperature is 22–24 °C, the monthly amplitude of temperature fluctuation in the year is low (4–5 °C), but the temperature range between day and night is very high, especially in the dry season, it reaches 10–12 °C, sometimes up to 15–16 °C. The lowest average temperature is 20–21 °C in January, the highest average temperature is in April and May is 24–26.5 °C, the highest is in Buon Dôn and Ea Soup (over 29 °C).

3.1.4.2   Rain Regime

Dak Lak has two distinct seasons, the rainy season normally starts from May then ends on November and accounts for 80–85% of the annual rainfall. The dry season is from November to April of the next year, and the rainfall only accounts for 15–20% of annual rainfall. However, the rain regime in Dak Lak is not consistent: In the Southwest region, the rainy season starts earlier than April. In contrast, the East usually starts later from late May to early June and ends in November–December.

3.1.4.3   Quantity of Rainy Day

The rainy season averages from 130 to 150 days/year, and there are approximately 15–20 rainy days/month and most are in July to September.

3.1.4.4   Evaporation Quantity

Evaporation capacity varies between regions from 850 to 1250 mm, and in the dry season a large amount of evaporation makes the soil have less moisture.

3.1.4.5   Air Humidity

The annual average humidity is 80–85%, the rainy season is excessively humid (the humidity is 90–93%), whereas in the dry season, there is a lack of moisture (the average humidity is only 70–75%), and the lowest the average humidity reaches is 40–45%.

3.1.4.6   Wind Regime

Dak Lak has two main wind directions corresponding to the monsoon circulation regime. In the dry season, the northeast monsoon is prevalent, and in the rainy season, it is mainly the southwest monsoon (Fig. 3.2).

Conclusion

The climate of Dak Lak is characterized by a tropical highland monsoon climate, strongly influenced by the monsoon tropical circulation regime. The topologic factor divides Dak Lak into the sub-regions with different climatic characteristics (Table 3.2).

BẢN ĐỒ LƯỢNG MƯA TRUNG BÌNH TRÊN NĂM TỈNH ĐẮK LẮK

Fig. 3.2 Map of distribution of average rainfall for five Dak Lak provinces

Table 3.2 Climate characteristics of some sub-regions of Dak Lak province

Station observed indexes	M'Đrăk (570 m)	Buôn Hồ (720 m)	Lăk (410 m)	Buôn Ma Thuột (480 m)
Average annual temperature (°C)	24.08	22.19	24.32	23.83
Average temperature in January (°C)	20.5	18.8	21.6	20.9
Average temperature in April (°C)	25.6	24.3	26.1	25.9
Total yearly temperature (°C)	8669	7988	8755	8019
Rainfall (mm/year)	2634	1739	2114	1949
Rain month quantity (>100 mm)	9	8	6	6
Annual average K2 humidity (%)	80	79	79	85
Evaporation (mm/năm)	1197	896	1245	1255
Dry index (X = S;A;D)	2;0;0	1;1;2	1;1;2	1;1;2

Source: Dak Lak Department of Natural Resources and Environment (2010)

3.1.5   Hydrological Regime

Dak Lak has an abundant river system, including two main river systems.

3.1.5.1   Serepok River System

The Serepôk river system flows in the West-Northwest direction into the Mekong river, with a basin area of 14,420 km², approximately 300 km of which runs through the province, and the riverbed is 100–150 m wide. The Serepok river has two main branches – the Krông Ana and Krông Nô rivers.

Krông Ana River   Krông Ana river is the main confluence of three major tributaries: Krông Buk, Krông Pach, and Krông Bông. The total river basin area is 3200 km², the main stream length is 215 km. The river flows in the east–west direction, along the river to the middle, downstream is the sour-soil marshes due to extended inundation. The slope of the large upstream tributaries ranges from 4 to 5%, the downstream distance in the Lăk region has a small slope of about 0.25%.

Krông Knô River   The Krông Knô river originates from the high mountains of over 2000 m running along the southern border of the province. After that, it changes direction to the north and merges with Krông Ana river in Buôn Đray and pours into Srepôk river here. The total area of the Krông Knô basin is 4620 km² and the main river is 56 km long, the average slope of the basin is 17.6%, the basin average height is 917 m, and the river network density is 0.86 km/km². The Krông Knô river includes two major tributaries, namely Dak Krông Kma, Dak Măng, and other small rivers.

3.1.5.2   River Systems Krông Năng and Hinh

This river system is located in the northwest of the province, with a catchment area in the territory of approximately 2880 km², with two main branches: Hinh river and Krông Năng river. In addition to the river system in Dak Lak province, there are many natural and artificial lakes such as Lake Lăk, Ea Kao, and Buôn Triêt.

3.2   Real Status of Land Use

According to the table of real land use in Dak Lak province, the land use structure has changed. The area of agricultural land and forestry land increased owing to the development promotion in agricultural production, and supportive policies for people receiving zoning and contracting for forest protection are increasing. Moreover, the master plan contributes to improving the efficiency of land use, reducing the area of unused land, and stimulating investors to participate in leasing land for productive development

3.2.1 Economy

The annual average growth of the total product value in the 2006–2010 period is evaluated at 12.34%, exceeding the plan target (planned goal: increases by 11–12% per year), in which the growth of construction industry reaches 20.33%; service reaches 22.14%; agriculture, forestry, and fishery reaches 6.03%.

In 2019, the economic growth rate of the province reached 9.23%; production value of agriculture, forestry, and fishery reached 41,755 billion VND (100% of the plan); industrial production value reached VND 16,500 billion; social investment capital mobilization reached 33,795 billion VND.

The total state budget revenue reached 6910 billion VND (equal to 101.5% of the plan, up 17.6% over the same period in 2018). Per capita income reached 49 million/person/year; the estimated poverty rate by the end of 2019 is 9.35%.

For the mission in 2020, Dak Lak province planned the targets: Total social product will reach about 62,500 billion VND; economic growth about 11%; per capita income will reach 54.6 million/person/year; ratio of poor households decrease from 4.36% to 4.49%; 40.1% (61/152 communes) meet the new rural standard (Fig. 3.3).

Fig. 3.3 Ban Me Thuot city at night

3.2.2   Society

Dak Lak province includes 13 districts, one city and one major town with 180 communal administrative units (13 wards, 13 towns, 154 communes). This is also the most populous province in the Central Highlands with nearly 1.9 million people. Dak Lak is the ninth largest administrative unit in Vietnam in terms of population with more than 47 ethnic groups living together, with 49.5% being Ê Dê, Nùng, Tày, M'Nông, H'mông, Thái. The rural population is approximately 1,400,000 people. In recent years, due to the continued implementation of urban embellishment, many new residential areas have been formed, so the population is still fluctuating between districts.

3.3   Survey, Analyze the Characteristics

The researched objectives are key areas, artificial reservoirs, hydro-electricity, and irrigation systems on rivers and fountains in the province for identification of flash flooding risk there.

3.3.1   Basic Concepts and Classification of Artificial Reservoirs

Dak Lak is a mountainous province in a highland monsoon tropical climate, with the outstanding features of: lower temperature compared to regions of the same latitude due to its average altitude of approximately 500 m above sea level. Dak Lak topography is strongly and deeply divided, high in the Northeast and Southwest, gradually sloping from East to West.

There are two main river systems: The Serepok river basin accounts for about 70% of the natural area and the Ba river basin accounts for about 20%, with hundreds of fountains located on the steep terrain and many rapids. Notably, the Serepôk River is the only river flowing backward from east to west.

The average annual rainfall of Dak Lak is from 1700 to 2000 mm. Dak Lak's economy primarily depends on perennial industrial crops such as coffee and rubber that require a lot of water for irrigation in the dry season. Moreover, the potential to develop hydro-electricity on the river and stream systems of Dak Lak province is very large, and it is the reason why Dak Lak is the province with the largest number of reservoirs in the country.

Until early 2011, the province had over 618 irrigational reservoirs and seven large and medium hydro-power reservoirs. Dak Lak province has favorable conditions for the development and construction of irrigation works that make a great contribution to the province's socio-economic development. However, they also pose the danger of creating tubing or flash floods and landslides, which have caused great damage to people and property of ethnic groups in recent years.

The reservoirs in Dak Lak province, if classified according to their origin, there are two types: natural lakes and artificial lakes; if classified according to the use purpose, there are hydro-electric lakes and irrigation lakes. In the irrigational lake system, depending on the volume capacity, including the typical lakes: large, medium and small; By investor, there are lakes built according to the planning of the specialized units and/or authorities at the same grade, in contrast, there are the lakes formed by the needs of each unit, organization, even due to demand of an individual household.

We now consider the role of each in the formation of flash floods:

Scientists classify flash floods as the following: first by congestion, second with steep slopes, third with mud, last is synthetic flash floods. In addition, there is also artificial flash floods – one of the most important forms of artificial flash floods is the collapse of artificial water reservoirs.

Artificial reservoir incidents as a type of flash flood are similar to those caused by flow blocks. Artificial water reservoir incidents are due to many causes: lack of planning, lack of basic investigation documents, shortcoming of design, construction, and management, as well as many other complex reasons. When the reservoir dam breaks, flood waves will cause flash floods similar to the type of congestive flash flooding. In general, flash floods that block streams or artificial reservoirs incidents often cause large flood waves, which are much more devastating than those of slope flash floods.

3.3.2   The Role of Lakes and Dams in the Formation of Flash Floods

Dak Lak has large natural lakes such as Lak and EaKao lake (Lak lake has the largest open surface area in Vietnam). In addition to the role of storing water for irrigation and people's lives, they also have the role of regulating the annual flow. They bring great benefits to the economic development of the province and play the role of micro-climatic regulation in a region; therefore, these lakes are not involved in the formation of flash and tubing floods.

A hydropower lake is an artificial lake built by humans, with the main purpose of generating electricity. A hydropower lake is designed and constructed according to a strict technical process, with specific operating procedures. Thus, this type of lake, if operated correctly, is also responsible for regulating annual flow, participating in flood regulation, and reducing the natural damage.

Contrarily for some reason, these lakes do not comply with the prefixed operating procedures – for example, the water flow to the lake cannot be forecasted; therefore, when there is heavy rain, the coming flow is too large, and to ensure safety for the lake, water must be discharged, which, along with rain on the concentrated zone or downstream of the heavy rain, is involved in the process of forming flash floods.

Medium and large irrigational lakes in Dak Lak such as Easup Lake and Low Krong Buk are specifically planned, built, and constructed in accordance with the design, operated by a specialized unit. Therefore, this lake type has the role of regulating the flow, hugely serving the province's economic development and the population that no one can negate. On the contrary, these lakes can repeat the same mistakes that cause flash floods like the abovementioned.

Dak Lak in the 1970s and 1980s of the twentieth century, due to the province's economic development goals, was also the national target of developing perennial industrial trees with high economic efficiency such as coffee and rubber. Rapid development has made Dak Lak the coffee capital of the country with the goal of leading Vietnam to become one of the major coffee exporting countries in the world; so, the surface cover of coffee and rubber increased rapidly.

Coffee is a plant that requires a very large water amount for irrigation in the dry season; this is also the reason for the formation of an irrigational reservoir dam for coffee of governmental units, even individual households. This type of lake is usually built on the basis of blocking flow on river systems, causing water levels to rise. Dams are mainly earthed dams.

During construction, depending on the investor's idea, sometimes professional drawings exist, construction is according to correct techniques; in contrast, sometimes the dams are built without technical drawings and construction plans (investment owner can be a farmer, a military entrepreneur, even a private household). These types of lakes and dams are often built on the tributaries of rivers and streams in the form of ladders, and this leads to incidents of artificial reservoir dam breaking causing flash floods. Flooding caused by artificial dam failure is a type of flash flood similar to congestive flood.

3.3.3   Damage of Flash Flood Caused by Accident in Artificial Reservoirs

With the deeply divided mountainous terrain, plus the human destruction of the natural environment in Dak Lak province in recent years and climate change, heavy rains cause tubing-flash floods, landslides, and damage to people and property, greatly affecting production and people's lives.

Some typical flash floods in Dak Lak: In 2001 due to the Buôn Bông dam breakage (in Ea Kao commune, Buôn Ma Thuôt city), 21 people died and many other properties washed away. In 2007, typhoon No.2 affected some areas with extremely heavy rainfall causing flash floods in Ea Kpam commune (Cư Mgar district), and flash floods washed away 3 people, 14 houses, 7 small traffic bridges, damaged 319 ha coffee, 63 hectares of cashew, 15 hectares of pepper, 512 hectares of rice, 339 hectares of corn and many other crops. In Xuan Phu commune (Krông Năng district), the flood swept away 9 people, of which 6 died and 3 people were saved, 31 houses collapsed or drifted, and many cultivated areas were damaged.

3.3.4   Research Results and Actual Situation of Reservoirs in Dak Lak Province

According to the statistics of the Board of Natural Disaster Prevention and Mitigation in Dak Lak Province, Dak Lak currently has over 600 reservoirs of all kinds, of which the quality status is as follows: about 25% of bad quality, 20% of good quality, the rest are of medium quality. The reservoirs of bad quality are commonly damaged as follows: The top-deck and body of the dam are degraded, trees have grown over the spillway, making the overflow ineffective, the dam body is cracked, the cultivators violate the safety corridor of the dam-lake, dam body is absorbed, dam surface and body are eroded, and the lakebed is deposited.

Drainage system of the lake: many lakes have no discharge valve or are damaged, the outlet gate is filled up. The current irrigation reservoirs do not have technical records and the operational process is rarely ensured with technical rules. A common situation is no regular maintenance due to lack of funding.

The irrigation dam-lake in Dak Lak province is built with high density along rivers and fountains of the Serepôk river system, in a ladder style. This type of construction can cause the dam breakage as a domino effect, resulting in a tubing-flash flood. These irrigational dam-lakes construction lacked the basic data of meteorology – hydrology; therefore, the normal water level rise, the overflow level, the catchment area, and the ability to concentrate water with the dam body design is sometimes not reasonable. Especially in the current global climate change context, extreme values happen with increasing frequency, so the risk of reservoir being unsafe is increasing.

3.4   Impact Factors Affecting Flash Flood Formation

Natural factors affecting the formation of flash floods in the province such as topography, geology, land resources, climate, and hydrology have been analyzed in detail in content No.1. This content only focuses on the human impacts on environmental factors that increase the risk of flash floods.

3.4.1   Land Environment

- **The situation of land use in Dak Lak**

According to the land statistic results of Dak Lak province in the 2010 almanac, Dak Lak has a total natural area of 1,312,573 hectares, including:

3.4.1.1   Agricultural Land

Total surface of agricultural land in 2010 was 533,404 ha. In the structure of agricultural land, perennial crop land has a surface of 314,884 ha, accounting for 59%, the total area of annual crop land in 2010 is 214,981 ha, accounting for 40.3%, the land surface for breeding grass is 1256 ha, accounting for 0.26%, water surface for aqua-culture has an area of 2283 ha, accounting for 0.44%.

3.4.1.2   Forest Land with Forests

The total surface of forest land is 599,908 ha, accounting for 45.71% of the natural land surface, of which natural forest is 568,142 ha, the surface of planted forest is only 31,766 ha and is planned to be divided into three forest types: productive forests, Protective forests, and specialized forests.

3.4.1.3   Resident Land

There are 14,368 ha, due to the custom as well as life style in many places, agricultural land mixed in residential land (mainly garden land).

3.4.1.4   Urban Land

In the only urban area Buon Ma Thuot, there are 37,718 ha, of which the residential land is 2318 ha. However, the urban area still has a sizable agricultural land surface of 26,315 ha (accounting for 69.86%). District towns have a large surface but have their own characteristics of the socio-economic center of the Central Highlands. Therefore, the real urban area of these towns accounts for a small proportion, the majority of the population is still works in agricultural production. The total urban residential land surface in 2010 was 2318 ha.

3.4.1.5   Specialized Land

The specialized land surface was 87,463 ha; generally, the area of special-use land in the whole province in recent years increased compared to the previous period (land for traffic and irrigation).

Table 3.3 Land use area of Dak Lak province in 2010

Distribution	Total surface (ha)	Type of land (ha)				
		Agriculture	Forestry	Specialized	Residence	Not yet used
Totality	1,312,537	533,404	599,908	87,463	14,368	77,394
Buon Me Thuoc	37,718	26,315	1005	6742	2318	1302
Ea Hleo	133,512	67,791	53,597	6856	1033	4235
Ea Súp	176,563	37,876	124,003	7460	574	6650
Krông Năng	61,479	42,607	8495	4944	1111	4322
Krông Búk	35,782	29,153	711	4940	597	381
Buôn Đôn	141,040	22,286	109,464	6102	557	2631
Cư M Gar	82,443	62,062	11,385	7269	1278	449
Ea Kar	103,747	51,289	37,859	7170	1407	6022
M'Drăk	133,628	33,449	77,252	7973	511	14,443
Krông Pắc	62,581	42,988	4248	8262	1561	5522
Krông Bông	125,749	27,319	80,405	3700	671	13,654
Krông Ana	35,609	24,218	5840	3571	618	1362
Lăk	125,604	18,713	84,724	6114	483	15,570
Cư Kuin	28,830	23,307	809	3040	888	786
Buôn Hồ town	28,252	23,995	111	3320	761	65

Source: Dak Lak Statistical Yearbook (2010)

3.4.1.6 Not Yet Used Land

Its surface is 77,394 ha. Most of the not yet used land of the province is scattered throughout high mountainous terrain, so it is very limited in use for agricultural purposes. In the long term, it is oriented to allocate about 60% of the not yet used land for forestry purposes (Table 3.3).

3.4.2 The Situation of Polluted Soil

According to research results of Mr. Nguyen Van Chien and the Institute of Agricultural Planning and Design, which have been converted to the FAO – UNESCO classification system, the whole Dak Lak province has the following soil types:

– Group of Alluvial soil (Fluvisoils).
– Group of Gley soil.
– Group of Peat soil (Histosoil).
– Group of Black soil (Andosoils và Luvisoils).
– Group of Basalt soil (Ferrasoils).
– Group of Alit mud on high mountains, symbol A (Haplic Alisoils – Alh).

Fig. 3.4 Tilling up on sloping land in Dak Lak province. (Source: IESEM 2011)

Dak Lak has a diverse land resource, but the maximal exploitation of the available potential has led to:

- Land fund and land quality are increasingly declining, degradation and infertility of cultivated land is quite common in mountainous areas due to inadequate use, watershed deforestation is causing erosion and strong wash-out. Mono-culture of some crops and using motorized means, in addition to faster and faster urbanization.
- The surface of natural forest tends to be narrowed, while the construction land, transportation, irrigation, and residential areas are increasing. This change of land area fund has a significant impact on the living environment of rural people in the province.
- In recent years, the surface cover of coffee, rubber, and other crops has successively increased, and the abuse of fertilizers and flora-protective chemicals to increase crop yields has degraded soil quality and negatively affected surface water and groundwater (Fig. 3.4).

3.4.3   The Human Impacts Increase the Risk of Flash Flood

- **Unreasonable land use**

3.4.3.1   Scientific Lack in Land Cultivation

Actual agriculture practices in the province such as highland field rice cultivation may increase the risk of flash floods. When reclaiming earth for cultivation, the forest cover disappears, causing the land to be affected by rain, causing erosion, wash-out and gradual degradation, with its rate fast or slow depending on the level of land cover, slope, surface runoff, and soil resilience.

The excessive expansion of highland-fields to the critical areas, planting to the top without keeping the trees there, reclaiming the hills by vegetation shaving has also led to soil degradation. Actions of cover trash cleaning, organic matter removal or weeding for production is also a cause of soil degradation in the province.

3.4.3.2 Indiscriminate Land Exploitation

Dak Lak is covered by many hills and mountains, has the advantage of land resources, and therefore has many land exploitative zones to serve residential projects such as leveling for industrial parks and road/ house construction, filling in lakes and ponds.

A lack of schematic planning has led to arbitrary exploitation, where easy and near are exploited and difficult and far are refused, depleting local land resources. Some areas that were originally mountainous were turned flat or even concave, causing ecological imbalance and instability of the base-ground; therefore, when heavy rains, drifted soil, and flash floods suddenly form, they will cause a lot of damage to people, property, and the local ecosystems.

- **The pressure of population growth**

Population growth in the province is mainly mechanical growth, concentrated in the city area, Buon Ma Thuot, and developing towns and remote communes are partly due to the large number of free migrants from the Northern and Central provinces into Dak Lak.

Population growth leads to an increase in subsistence and other necessities, while natural resources are limited, especially land for agriculture and residence. The inevitable consequence is to expand the agricultural land into forest land, causing deforestation and reclamation throughout the province, increasing bare land and hills, increasing soil erosion, thereby increasing the risk of tubing–flash floods.

- **The impact of urbanization and industrialization**

The expansion of urban space leads to the occupation of agricultural land and other land types for urban construction. Land is fully exploited, the proportion of trees and water surface in urban areas and industrial zones decreases, and the concrete covering affects ground-water sources, causing soil pollution.

The process of urbanization and industrialization is often linear with the input of natural resources to serve them: trees are cut for use as wood, paper or firewood, but the forest is not compensatively planted.

Urbanization and population growth has led to poor infrastructure, including the illegal construction of houses on canals, encroachment on river-beds, and disposal of domestic wastes into canals thence making a blocked deposition, increasing the damage level while reducing flood drainage during heavy rain and floods.

Urban construction often causes the loss of vast fields, ponds, lakes, lagoons originally used to store the flood water in a natural way, making flash floods worse. Therefore, the urbanization process, if not under the law on compliance, under regulations on zoning plan of land use will increase the risk of flash floods in the urban development process.

Industrial development led to the increased exploitation of natural resources. Therefore, without appropriate planning and management measures for resource exploitation and use, the resources will be degraded and depleted rapidly.

- **The impact of deforestation**

According to the report of Dak Lak Provincial People's Committee, the total surface of forest and forestry land in the area is more than 720,000 ha. In which, the forested land area is over 526,000 ha, the forest coverage is over 39.3% (including rubber trees). The forest was allocated to 15 forestry companies, 7 specialized forest management boards, 4 management boards for protective forests, 69 enterprises with functions of management, production and business organization, forest zoning and protection and a part assigned to the People's Committees of districts and communes for management,

According to statistics of Dak Lak Provincial Forester Department, in 2017 alone, functional forces detected and handled 1407 cases of forest law violations, confiscated 2441.7 m^3 of timber and 717 vehicles of all kinds, and collections paid to the state budget were more than 18.8 billion VND. Compared to 2016, the number of forest law violations decreased by 217 cases; although the number of cases decreased, large-scale and daring deforestation cases still occurred.

One of the reasons leading to deforestation is that local authorities with forests have not fully implemented their assigned responsibilities in forest management and protection. Some officials and civil servants in the forest protection force still lack responsibility, even cover up and support the illegal loggers, so gaps remain for illegal businesses to take advantage of violations.

Deforestation changes the plan and use, the vegetation, and forests turn into grasslands. Trees are capable of retaining water as well as minimizing land loss. The amount of flood water in an area with plenty of trees is less than the flood water amount in a barren area. Therefore, deforestation will increase water levels in the downstream areas.

Deforestation causes soil degradation, infertility, oxidation, and inert rock/gravel erosion. Deforestation destroys a wide range of forest ecological functions, such as water regulation and protection, aerial purification, and climate moderation. In the dry season, hot and dry winds are raging, dust storms sweep away a lot of fertile soil, and the water resources are exhausted, so the land is barren and hard. Meanwhile, in the rainy season, soil erosion, landslides, flash floods sweep away crops and property.

Substandard logging also produces large amounts of sediment and can affect local flow styles, particularly by increased runoff from gathering yards, slide-ways, and logging tracks.

- **Land use practices**

Dak Lak today is a cultural exchange place of many ethnic and local groups. Dak Lak province has 47 ethnic groups, accounting for over 40% of the total province population, which is 1734 million people. Of which Ede, M'nong, and J'rai are the main indigenous or local ethnic groups, while other ethnic groups have migrated over the past 30 years, such as Tay, Nung, M'Nông, Dao, Thai, and H'Mong. Most ethnic groups still retain their own cultural heritages, forming a color featured array in the entire cultural life of the Vietnamese Plateau or the Central Highlands.

These ethnic groups have the practice of slash-and-burn cultivation according to the method of reclamation, burning up the slope fields for planting rice and various kinds of food crops with the purpose of short-term land use, intermittent, timely vacancy, then from mobile cultivation to nomadic life. This form of cultivation is strenuous, but the productivity is low, life is precarious, and depends entirely on nature. This cultivation form causes huge environmental damage and consequently increases the area of barren land, fallow, erosion, and risk of flash floods.

3.4.4   Causes of Biodiversity Degradation

As we know, population growth has led to more human pressure on natural resources and the environment, in which biodiversity accounts for a very important part. It is also the cause of the decrease in Dak Lak province bio-diversity in recent years:

- Exploitation of forest resources and specialties has greatly affected ecological balance, and the function of forest protection has been impaired.
- Field burning and wildfires. The slash-and-burn method is a long-standing custom of ethnic minorities in mountainous and highland areas, thence forest fires are inevitable. This is one of the reasons for the reduction of bio-diversity in the province.
- Planting and exploiting the industrial and agriculture crops: Before the leverage of the market economy, coffee prices surged in the 1990s, then a large forest area had to give way to coffee, pepper, and cotton.
- Free migration: This is a hot problem for Dak Lak like some Central Highlands provinces, most of the people who migrate freely choose upstream and forbidden forests to live and work. Because here is the old forest, good land, and many resources from the forest are profitable to people's lives.
- Impact of chemical toxins: The heavy consequences of war toxins destroy organisms and ecology in many areas, cause heavy environmental pollution, and negatively impact people's lives.

– Because of the inadequate investment in forest protection, forest owners do not really care about their forest. During the exploitations, they have not obeyed the procedures for cutting, sanitation, and restoration after exploitation.

3.4.5 Actual Status and Evolution of Bio-Diversity Degradation

Dak Lak is one of 12 biodiversity centers in Vietnam. Forest and agricultural ecosystems are very diverse and rich in a number of species. The Dak Lak forest has many endemic flora and fauna, especially the typical dipterocarp forest ecosystem of the country (Fig. 3.5).

3.4.5.1 Forest Vegetation

The tropical forest is often green, develops mainly on high mountainous terrain, has thick soil layers, streams, and watersheds comprising a multilayered ecosystem, with many rare and precious trees such as Rosewood, Xylocarpa, Hopea Pierrei, Foklenia, Magnolia, and Kessiya Pine and precious pharmaceutical herbs are being conserved and developed, with good coverage, a thick carpet layer, and porous soil.

Fig. 3.5 Central Highland dipterocarp forest

Table 3.4 Plant composition in some conservation areas

Investigational position	Number of plant family	Number of plant species	Number of ornamental plants	Number of pharmaceutical plants
Nam Kar natural conservation park	149	587	78	382
LăkLake – historic/cultural/environmental zone	116	548	14	264
Chư Yang Sin national park	56	130	10	30
Chư Hoa area	110	353	14	53

Source: Construction project of conserved area

Mixed forest of Bamboo, Neohouzeaua and Wood: This is a forest with the main composition of Bamboo, Neohouzeaua, Rattan, and Dipterocarpa trees.

Regenerative young forests and shrubs are the results of forest exploitation over many years; Vegetation is mainly Dipterocarpa, Legume, Oval, Chestnut, Anthela, and Reed from 2 to 15 m height. The young regenerating forests and shrubs are evenly distributed throughout the province.

Natural grass cover: distributed in many different terrains, together with fallow fields forming a large grassland (Table 3.4).

3.4.5.2   Forest Fauna

In the above natural reserves, only a few areas have formed the organizational structure, management and operation such as: Yôk Dôn National Park, and other Natural Conservation Zones of Lak Lake, Nam Kar, Ea So, and Chư Yang Sin, the rest have not yet established a managing unit, and have no research, protection, and development activities (Table 3.5).

3.4.5.3   Evolution of Biodiversity Degradation

Nowadays, Dak Lak is one of the most notable areas for biodiversity and endemic species, including many rare and precious species such as elephant, tiger, leopard, and bison. Forest fauna has 228 species belonging to 26 families, 11 orders; wild birds have 598 species, 46 families, 18 orders; crawlers 129 species of 12 families, 3 orders; frogs & amphibians 79 species, 5 families, 2 orders; fish have 98 species.

The distribution of the afore-mentioned flora and fauna species is quite wide, mostly in the high mountainous tropical, subtropical, and semi-tropical vegetation areas, and with the broad-leaf evergreen tropical forest, dipterocarp or semi-deciduous forest.

Table 3.5 Typical Fauna composition in some conservation zones

Specialized Forest	Reptiles			Frogs (Amphibia)			Birds			Animals		
	Species	Family	Order	Species	Family	Order	Species	Family	Order	Species	Family	Order
Chư Yang Sin	29	11	2	15	4	1	130	43	16	49	26	11
Hồ Lắk	26	10	3	17	4	1	132	42	1	61	25	10
Nam Kar	34	12	3	16	4	1	140	43	17	56	24	9
Yok Đôn	40	12	2	31	5	2	196	46	18	62	26	11

Source: Construction projects for conserved areas

The fauna is rich in quantities of species and individuals. Abundant habitats could exist to facilitate the recovery and development of fauna, of wetland bio-diversity, in agriculture, afforestation, and in landscape. However, in recent years, due to many different reasons, Dak Lak natural resources have experienced many changes such as exploitation of timber and specialties, slash-and-burn cultivation, land clearing for agricultural and forestry, illegal hunting, and free migration.

3.5   Assessment on the Actual Situation

3.5.1   Overview of Flash Floods in Dak Lak

According to statistics, the province almost annually has flash floods, landslides in the rainy season, and in some places, flash floods happen many times. Flash floods often arise suddenly, in a narrow range but are very fierce and often cause great losses and serious damage.

To prevent flash floods, and minimize the damage caused by flash floods and landslides, we need to review and evaluate the risk of flash floods in areas with steep terrain and lowland areas, and water reservoir projects to develop the appropriate solutions to prevent flash floods and landslides, and to flexibly and effectively apply these solutions in practice.

Assessment of damage caused by flash floods and landslides is based on aggregate collection and analysis of 10-year data chain (2000–2009) about flash floods that occurred and their damage for socio-economy.

3.5.2   Direct Losses

In the past 10 years, flash floods and landslides in the province have killed 39 people, hundreds of households lost their homes and all assets; thousands of hectares of crops were damaged and hundreds of hectares of productive land were buried and could not be used. Many infrastructure projects such as traffic, irrigation works, schools, and medical stations were damaged. Economic losses amounted to hundreds of billions vnđ.

These are the big damages; moreover, those damages are mainly concentrated in the highlands and remote areas where the education level and economy are low. Compared with river floods, the economic damage caused by flash floods is lower, the impact is narrower, but flash floods cause great loss of life, especially people in remote areas. Most of them are ethnic minorities and poor households, subbject to the current preferential care policies of the State.

3.5.3   Indirect and Longtime Losses

Considering a specific area, flash floods and landslides not only cause serious consequences for a certain area at the existing time, but often the consequences remain much later. Finance for overcoming the consequences of flash floods and landslides in order to stabilize people's life and production is an urgent concern right after the disaster. A wide range of problems need to be addressed such as providing food for hunger, treating diseases, repairing or rebuilding houses.

To solve those urgent problems requires a large amount of funding, most of which overpassed the localities' capacity, so the "4 on-site" motto is difficult. For areas at high risk of flash floods and landslides, there is a need to have a plan to migrate out of the danger zone, addressing this task requires coordination of many branches and a rather large budget for implementation.

Flash floods and landslides often severely destroy the infrastructure works, bury or erode the large areas of agricultural land and crops, leading to disruption of agricultural production, reducing yield and food productivity. In some places, fields were eroded or buried by soil and rock from 1 to 2 m, resulting in loss of cultivated areas. These may lead to an increase in deforestation for land exploitation or an increase in deforestation activities in search of other resources to replace production.

On the other hand, because the majority of areas affected by flash floods are remote areas, when heavy rains cause flash floods, they also cause landslides that block traffic, making it difficult for rescuers to access areas after natural disasters. These real shortcomings have hindered the efforts of communities inside and outside the affected areas to self-overcome and carry out relief, rescue work to stabilize housing and production.

Environmental degradation in the flash floods area is inevitable. Drinking and domestic water sources are polluted, land is washed away, fields are buried, carpet covers are damaged, and the ecological balance of the area may be disrupted. Restoring the environment after flash floods and landslides requires a lot of effort over a long period and requires cooperation from many industries to create a clean and stable environment.

3.5.4   Research Results and Reality Review in Seven Districts

3.5.4.1   District Ea Hleo

Situated on the North of the province, the topography is strongly divided, the average annual rainfall is 1630 mm; the natural area is 133,512 ha, of which 57,041 ha is agricultural land, 61,062 ha is forestry land; population more than 120,000 people. There are 37 irrigation works, mainly small and medium reservoirs. In recent years, rains with very high intensity often appear, causing flash floods and landslides. On August 2, 2007, an extremely intense rain occurred, causing flash floods and landslides in some areas of the district. The followings are investigation results of areas affected by flash floods and landslides.

Ea Drăng town: Quarter 3 area was flooded on August 2, 2007, landslide on the hillside, killing 1 person and 1 house collapsed, 5 other houses and some properties were buried and severely damage.

Ea Tir commune: The area of village 3 was affected by flash floods and landslides, 9 houses were severely damaged, 25 houses were flooded and covered with mud and rock, drifting 1 temporary bridge eroded 2 km of traffic, eroded Ea Ru flood discharge and caused much other property damage.

Ea Khal commune: The area of village 6 and village 2 are the places affected by flash floods and landslides, damaged 8 houses and 1 school, flooded 20 households, drifted 4 temporary bridges, and damaged some assets, including other products such as crops, livestock, materials, food, and household goods.

Cư Amung Commune: Villages of 4, 7 were affected by flash floods and landslides, 4 houses collapsed, 26 houses were flooded and covered with mud and rocks; damaged 1 irrigation work. Much other property was damaged such as materials, crops, livestock, and home tools.

Ea Wy Commune: Villages of 4, 9 were affected by flash floods, killing 1 person and injuring 1 person; flooded, damaged 36 houses and 2 schools; drifting 01 bridge, 02 culverts and washed away some other property such as plants, animals, food, materials, and home tools.

Cư Môk Commune: Hamlet 1 is the area affected by flash floods and landslides, 19 houses were flooded and filled with mud; 200 m of provincial road 19B was eroded; damaged one irrigation work. Many other assets such as crops, animals, food, materials and home tools were washed away.

In addition, some areas of Ea Hleo and Ea Son communes are also at high risk of flash floods and landslides. The flash floods occurring in the district are mainly sloping floods. Most people do not pay attention to the danger of flash floods and landslides; most government officials have little experience in preventing these dangerous disasters.

3.5.4.2 District Krông Buk

Complex terrain, with many steep slopes. The average annual rainfall is 1572 mm; the natural area is 35,837 ha, of which 28,683 ha is agricultural land, 637 ha is forestry land; population of 60,000 people. There are 29 irrigation works, the most common being small and medium lakes with inter-lake relationships. In recent years, there have been two flash floods and landslides (2003 and 2007) in Chư Pông and Chư Kpô communes. The followings are survey results of areas affected by flash floods and landslides.

Cư Pông Commune: This area has many steep slopes, poor carpet cover; Ea Klôk stream is narrow but the streambed has a steep slope. There were 2 villages affected by flash floods, the event damaged 4 houses, 1 small irrigation project, washed away 2 wooden bridges, eroded many village roads, and washed away some other assets such as materials and furniture.

Ea Sin Commune: There was 1 village affected by flash floods and landslides, damaging many coffee areas, drifting a temporary bridge, and eroding many sections of rural roads.

Cư Nê Commune: There were 2 villages affected by flash floods and landslides, damaged many areas of coffee, broke many small lakes and damaged some other assets such as crops, livestock, materials, food, and home appliances. In the villages of Mui and Drao in 2001, a 800 m-long and 4 m-deep crack caused a landslide in many coffee gardens.

Chư Kpô Commune: In 2007, the residential area of Chư Kpô Rubber plantation was affected by flash floods and landslides, killing 1 person, damaged many houses, broke an irrigation dam combined with roads, drifting 1 small market, damaged many coffee areas, and washed away some other assets such as materials, animals, and home tools.

The flash floods that occur in the district are mainly flash floods on steep slopes.

3.5.4.3 District Krông Năng

Located in the northeast of the province, a region with many streams, the slope of the stream bed and the hill/mountain sides is quite large, the average annual rainfall is 1800–2200 mm/year; the natural surface is 61,479 ha, of which 38,234 ha is agricultural land, 6309 ha is forestry land; population of about 120,000 people. There are 59 irrigation works, mainly reservoirs. In 2007, flash floods swept away 9 people, of which 4 died, 2 were injured; collapsed and seriously damaged 14 houses; damaged a lot of crop surface. Phu Xuan, Ea Dah, and Cư Klông communes and towns are areas at risk of flash floods and landslides.

Phu Xuan Commune: Xuan Thai II village was affected by flash floods and landslides. The 2007 flood swept away 9 people, of which 6 were dead and missing, 3 were injured; 14 houses collapsed, drifted and were badly damaged, eroded many roads and washed away many other assets such as crops, food, materials, and home tools.

Ea Dăk commune: is a remote commune, affected by flash floods and landslides, with damage to 10 houses and some other assets such as crops, livestock, materials, food, and home appliances.

Krông Năng Town: Binh Minh Village was affected by flash floods and landslides, with 7 houses damaged. Many other assets such as materials, crops, livestock and home tools were washed away.

Cư Klông Commune: affected by flash flood, killing 2 people, damaged some houses and swept away some other assets such as crops, animals, food, materials, and household tools.

In addition, some areas of Tam Giang, Dliêya, and Ea Tan communes are also at high risk of flash floods and landslides. The flash floods occurring in the district are mainly sloping floods.

3.5.4.4   District M'Drăc

Located in the east of the province, with mountain slope, river bed, and steep stream; the vegetation was strongly exploited, the average annual rainfall is 2110.7 mm; the natural area is 133,628 ha, of which 24,693 ha is agricultural land, 67,704 ha is forestry land; population of more than 65,000 people. There are 41 irrigation works, mainly small and medium reservoirs. In recent years, there have been a number of small-scale flash floods and landslides.

Damage caused by flash floods and landslides in M'Drăc district over the past years has mainly been to houses, roads, irrigation works, and crops. Areas prone to flash floods and landslides include: Cư Króa, Cư Mta, Krông Jin, Ea Trang, Cư San, and Krông Á. The following are the investigation results of areas affected by flash floods and landslides:

Cư Króa Commune: There were 2 villages with 50 households affected by flash floods and landslides, 7 houses damaged, 1 traffic bridge swept away, and some other assets such as cattle, poultry, materials, and home appliances.

Cư Mta Commune: There was 1 village with 134 households affected by flash floods and landslides, damaging 59 houses of which 8 houses collapsed and completely drifted; eroding 2 km of roads; drifting 72 Cows and 3 tons of rice seed and many other properties.

Krông Jin commune: There was 1 village and 3 ethnic minority villages with 20 households affected by flash floods and landslides, damaged some houses and other assets such as crops, animals, materials, food, and home appliances.

Ea Trang Commune: There were 3 villages with 20 households affected by flash floods and landslides, damaging houses and swept away many assets such as materials, crops, animals, and home appliances.

Cư San Commune: There were 8 villages with 22 households affected by flash floods and landslides, damaging houses and swept away many assets such as materials, crops, animals, and home appliances.

Town: There were 3 villages with 120 households affected by flash floods and landslides, damaging some houses and other assets such as crops, animals, food, materials, and home appliances were swept away.

In addition, some areas of Krông A commune are also at high risk of flash floods and landslides. The flash floods occurring in the district are diverse in terms of their causes, with flash floods on steep slopes, congestion-blasting floods, and mixed floods. Most local people do not pay attention to the danger of flash floods and landslides. Regional government officials have little experience in preventing these dangerous disasters.

3.5.4.5 District Cư Mgar

The average rainfall is 1920 mm/year; natural area 82,443 ha, of which 61,385 ha is agricultural land, 8937 ha is forestry land; population of 165,000 people. There are 47 irrigative works, mainly reservoirs, many lakes related to inter-lake. In the past 10 years, flash floods and landslides of this district caused 9 deaths, collapsed and severely damaged 52 houses, damaged 2300 hectares of crops. The flash flood on August 2, 2007 killed 3 people, drifted 14 houses, 7 traffic bridges, broke 2 irrigation dams, and damaged many cultivated areas.

Cư Mgar district has many reservoirs, including several inter-lake systems. The work was built in many periods, up to now, and a number of projects have deteriorated and safety is not ensured. In addition, the side-slopes are poorly covered, due to the influence of morphology and the considerable amount of rainfall in this area, especially in recent years, there have been many extremely high intensity rains. Areas in Cư Mgar district at high risk of flash floods and landslides include Ea Kiêt, Ea Tar, Ea Mdroh, Ea Hđing, Ea Kpam, and Cư M'gar communes. The following are investigation results of areas affected by flash floods and landslides.

Ea Kiet Commune: There was 1 village with 15 households affected by flash floods, 5 houses damaged, 1 traffic bridge swept away and some other assets such as cattle, poultry, materials, and home appliances.

Ea Tar commune: There were 3 villages and 34 households affected by flash floods and landslides, damaging 7 houses; eroding some roads and drifting away many other assets such as poultry, materials, and home appliances.

Ea Mdroh Commune: There were 4 villages with 32 households affected by flash floods and landslides, which damaged some houses and some other assets such as crops, animals, materials, food, and home appliances.

Ea Hđing commune: There were 3 villages with 25 households affected by flash floods and landslides, damaging 25 houses and swept away many other assets such as materials, crops, animals, and home appliances.

Ea Kpam Commune: There were 2 villages with 12 households affected by flash floods and landslides, causing 5 deaths; damaged some homes and washed away many other assets such as materials, plants, animals, and home appliances.

Cư M'gar commune: There were 2 villages and 15 households affected by floods and landslides, 1 person dead; damaged some homes and many other assets such as crops, livestock, food, materials, and home appliances were washed away.

The flash floods occurring in the district are diverse in their formation causes, with flash floods on steep slopes, congestion-blasting floods, and mixed floods. Most local people do not pay attention to the danger of flash floods and landslides. Most government officials have little experience in preventing these dangerous disasters.

3.5.4.6 District Lăk

As a southern upper source of the Krông Na river, the area has many steep slopes and large and small streams, and the average annual rainfall is 2020 mm; the natural area is 125,605 ha, of which 17,677 ha is agricultural land, 82,365 ha is forestry land; population of 60,000 people; forest coverage reaches 62.3%. There are 28 irrigation works, the most popular being the reservoir. This is a low-lying area with frequent floods and inundation every year in the rainy season.

Besides the fluvial floods, this area is also a common area of frequent flash floods and landslides due to the divided topography with many steep slopes on hillsides and streambeds. The communes that have experienced and are at high risk of flash floods and landslides are Yang Tao, Bông Krang, Krông Nô, Dak Liêng, Buôn Tría, and Ea Rbin. Here are investigation results of areas affected by flash floods and landslides.

Yang Tao commune: There were 3 villages with 219 households affected by flash floods and landslides, some houses were damaged, 1 bridge for traffic and other assets such as cattle, poultry, materials, and furniture was washed away.

Bông Krăng commune: There were 4 villages with 250 households affected by flash floods and landslides, causing 24 damaged houses; 3 km erosion of roads; washed away many other assets such as livestock, poultry, food, materials, and home appliances.

Krông Nô commune: There were 6 villages with 400 households affected by flash floods and landslides, which damaged a number of houses and some other assets such as crops, animals, food, materials, and home appliances.

Dak Liêng Commune: There were 9 villages with 301 households affected by flash floods and landslides, damaging houses and swept away many other assets such as materials, crops, livestock, and home appliances.

Buôn Tría Commune: There were 5 villages with 347 households affected by flash floods and landslides, damaging houses and swept away many assets such as materials, plants, animals, and home appliances.

Ea Rbin: There were 3 villages with 162 households affected by flash floods and landslides, damaging some houses and other assets such as crops, livestock, food, materials, and home appliances were washed away.

3.5.4.7 District Krông Bông

This is Krông Bông river basin with many branches of big/small tributaries and streams. The terrain is strongly divided, many slope areas have very poor cover (bare hills), and the streambed has a steep slope. The average annual rainfall is 1800 mm; natural area is 125,748 ha, of which 25,617 ha is agricultural land, 77,980 ha is forestry land; population (2009) of 90,000 people; forest coverage reaches 20–30%. There are 28 small and medium irrigative works.

Besides the natural factors of topography and land cover, this area also has many degraded small reservoirs that pose a safety risk. Krông Bông River passes through the territory of 11/14 communes and towns in the rainy season and is often affected by riverbank erosion. Communes at risk of flash floods include Cư Drăm, Cư Pui, Jang Mao, Hoa Le, Ea Trul, and Hoa Tan. The followings are investigation results of areas affected by flash floods and landslides.

Yang Mao Commune: There were 2 villages with 25 households affected by flash floods and landslides. The disasters damaged 11 houses, eroded many sections of rural roads and swept away some assets such as livestock, poultry, materials, and home appliances.

Cư Drăm Commune: There were 3 villages with 34 households affected by flash floods and landslides, causing damage to 9 houses; accretion of many rice fields; drifting some livestock, materials, rice seeds, and household appliances.

Hoa Phong commune: There were 2 villages and 1 ethnic minority village with 32 households affected by flash floods and landslides, which damaged a number of houses and some other assets such as plants, animals. Materials, food and home appliances.

Cư Kty Commune: There was 1 village with 15 households affected by flash floods and landslides, which damaged 6 houses and washed away many assets such as materials, crops, animals, and household appliances.

Hoa Le Commune: There were 3 villages with 45 households affected by flash floods and landslides, which damaged houses and swept away many assets such as materials, crops, animals, and home appliances.

3.6   Recommendations on Framework Strategy

Combining the method of system analysis and experts' consultancy to build solutions to prevent and reduce damage caused by flash floods in Dak Lak province:

– *Group of solutions for mechanisms and policies:* research and propose the rational solutions for schematic planning, land management and use (forest land, agricultural land, residential land, infrastructure land, etc.), solutions of management and residential relocation….

– *Group of solutions for science and technology:* propose the structural measures (cultivation techniques on sloping land, hydrological monitoring system, communication and flash flood warning system, etc.).

– *Group of solutions for building and developing resources.*

– *Group of educational solutions to improve environmental awareness:* to help people actively participate in the prevention, mitigation, and recovery of flash flood damages.

  • Education and propaganda in key areas at risk of flash floods.
  • Permanent education in schools.
  • Community education.

– *Propose the unification on command unit, implementing agency, avoiding overlapping in the implementation of solutions.*

– *Other support solutions.*

Chapter 4
Establish the Databases for the Map of Flash Flood Risk (Case Study: Dak Lak)

Abstract This chapter collects data for flash flood maps. Based on the assessment of the current flash flood situation at the statistical locations throughout Dak Lak province, a database for the development of flash flood risk maps is established. This process analyzes the potential occurrence of flash floods based on closely related factors, such as rain, topography, river network, soil, and actual land use, combined with analysis software for collected data to build a flash flood risk map.

The purpose is to make a database with all the parameters of weather, hydrology, area, flood volume, elevation, slope, soil permeability, vegetation layer, and farming situation as the basis for making flash flood risk zoning maps at the scale of 1/10,000 for the whole province of Dak Lak and the scale of 1/25,000 for high-risk areas, and at the same time, systemizing the decentralize groups of risk factors for flash flood formation in the area.

Keywords Map databases · Data information · Zoning map · Elevation model

4.1 Assessment of Flash Flood Risk on Dak Lak

As part of information analysis on the socio-economic development of each region in the coming years, the recent occurrences of flash floods and landslides have been noted, using topographic maps for research. Research shows that in the seven districts of the province being investigated and reviewed, there are many areas prone to flash floods and landslides.

Buffer surface factors such as vegetation carpet and surface geology are easy to change, especially in recent years, after the exploitation of primarily natural forests, the number of afforestation surfaces is not enough to compensate for the exploited forest area. Reclaiming production land, developing settlements of free migrants not according to planning, along with unsafe irrigation work systems that have dam failure during heavy rains, all contribute to causing flash floods (such as broken Buôn Bông lake, Ea Kao commune, Buon Ma Thuot city, and Chư Kpô dam in Krông Buk district).

© The Author(s), under exclusive license to Springer Nature Switzerland AG 2022
L. H. Ba et al., *Flash Floods in Vietnam*,
https://doi.org/10.1007/978-3-031-10532-6_4

In particular, some new settlements located in high-risk areas exposed to flash floods are principal factors for an increased risk.

4.1.1 Potential Flash Flood Analysis on Base of Closely Related Factors

The formation of flash floods is the result of a combined impact of many factors:

- **Rain and weather patterns cause heavy rain**

Dak Lak has a deep seasonal differentiation that results in a large difference between the received water amount during the wet and dry seasons. Over 80% of the water that Dak Lak receives annually is concentrated in the rainy season months (from May to October). In the dry season, the rainfall is only approximately 15–20%, in some places only 8–15%. Therefore, it leads to scarcity, water shortages, even severe droughts in the dry season and inundation, flash floods in the rainy season that cause great damage to people and property.

During the flood season, Dak Lak is also affected by storms and tropical depressions. These storms, when entering Dak Lak province, have weakened into low pressure or a zonal-one but still produce heavy rain on a large scale. The maximum rainfall due to the influence of a storm can reach 150–500 mm, of which the maximum daily rainfall reaches 100–200 mm. Stormy rain occurs 1–3 days before the storm hits and continues for several days after. The frequency of large floods causing inundation caused by storm effects combined with the highly active southwest monsoon or the intensified cold air is 55–65%, particularly flash floods and landslides are 20–30%.

The activities of storms and tropical depressions to the formation of normal floods and flash floods in Dak Lak can be divided into two periods:

- *During the period from July to September,* storms and tropical depressions have little direct impact, but mostly due to the active southwest monsoon causing continuous rains for many days. In which, in the days when the storm crosses the central coast, usually the rainy days in Dak Lak have an increase in three features: quantity, intensity and rainfall size, so this is the time when there are big floods on the river system in districts such as Ea Soup, Buôn Dôn, and Cư M'gar. At this time of year, the risk of flash floods and landslides is also the highest.
- *From mid-September to the end of the season,* storm activities and tropical depressions tend to move south, so the affectability of floods in Dak Lak is higher. This is the period when rivers and streams have the highest water level and flow volume in the year. The damage caused by storms will be especially serious when it is affected by two consecutive storms in a week. The land is full of water, the water level of rivers and streams is high, low-lying areas have not been drained in time due to the first storm, the second storm has entered, in addition to the harm caused by the storm, the stormy rainfall causes the water level of rivers and streams to rise quickly, the flood area will expand, and floods will become very dangerous.

- **The terrain is steep, strongly divided**

Dak Lak topography is quite diverse and complex, with hills and mountains, the plains and valleys are quite steep, strongly divided by a system of rivers and streams that are relatively thick like the tree branches. These are favorable conditions for the formation of flash floods.

- **System of rivers and fountains**

The river network in the basin is often characterized by these five basic features:

- River level (grade)
- Branch length (km)
- River density (quantity of rivers/km^2)
- River network density (km/km^2)
- Runoff length (km)

The impact study of the river network on flash floods mainly focuses on the fourth characteristic: river network density. The river network density has an influence on the flow concentration process. A good drainage place has a thick network of streams, short runoff length, quickly concentrated surface flow, and high flood peak.

The system of rivers and fountains in the province is quite plentiful, evenly distributed, but due to the steep terrain, the water retaining capacity is low, the small fountains almost have no water in the dry season, so their water level is usually very low. The rivers have some common features as follows:

- Short river streams, mountainous streams with high riverbed slope in some places up to 40–50% (Dak Krông Kma stream). Streams in flat plains with great bends.
- River stream flow on complicated terrains, mainly originating from hilly and mountainous areas running through low-lying plain.
- Riverbed is not stable, on many river sections, the erosion phenomenon happens quite strongly.
- In the dry season, the water amount is poor, but during the rainy season, floods happen very fiercely.

- **Land and land use**

Soil is a principal factor of the buffer surface. Surveying the factor groups that cause flash floods, it is noted that rain is a necessary condition, so buffer surface is a sufficient condition.

The buffer condition strongly dominates the flood formation process. The buffer surface affects the loss amount of flood flow. Flood flow losses include permeation, depressions, stoppage by vegetative mulch, and evaporation. Permeability plays the most important role therein mainly determined by the soil. Soil properties also determine the composition of solid matter in the flood flow. Thus, soil directly affects the formation of flash floods in both phases: solid and liquid phases.

Dak Lak province groups have many favorable features for the flash flood formation:

- Having many steep terrains, strongly divided.
- The soil layer is thin, with heavy mechanical components, easily saturated in surface water or light mechanical components, easily permeable but weakly linked, easy to slip, erosion.
- Easy to become vital in case of inappropriate land use.

Buffer surface conditions are favorable for erosion, leaching, and landslides during heavy rain and floods. These conditions often converge in small mountainous basins, thick river networks, steep slopes, short rivers, narrow valleys, where over-exploited, residential/urban areas, and river works that obstruct flow. Especially at the beginning of the rainy season, or when there is less rain, but the soil is weathered strongly, the favorable conditions for erosion and washing-out are more evident.

This is the most important cause of flash floods during heavy rain conditions. If the above-said buffer conditions cannot be converged, even with heavy rain, even the biggest in the basin, flash floods have not occurred in river basins.

4.1.2   Study to Establish Zoning Map of Flash Flood Risk

Map scale 1/100,000 for Dak Lak province and scale 1/25,000 for two main areas.

(a) **Information about data and software**

Data Information
- Daily rainfall data (mm):
- From 1980 to 2019
- Source: Dak Lak Meteorology and Hydrology Center
- SPOT4 image data:

  - Data source: Southern center of remote sensing technology
  - Data Model: Raster
  - Data type: Byte
  - Reference system: VN2000
  - Reference point-mark: WGS84
  - Unit of reference: Meters
  - File type: Binary
  - X resolution: 20 m
  - Y resolution: 20 m
  - Year of shoot: 2010
  - Release year: 2010
  - Format Type: GeoTIFF
  - Sensor: HRVIR

– Land distribution map and land use status quo map of Dak Lak province

  • Source: Department of Natural Resources and Environment of Dak Lak
    province
  • Reference System: VN2000
  • Year of implementation: 2010

– Topographic map of Dak Lak province

  • Source: Vietnam General Department of Land Administration
  • Reference System: VN2000
  • Year of implementation: 2005

Software Information
• Professional
• Version: 10.0
• Source: MapInfo company (now Pitney Bowes)
– Envi

  • Version: 4.7
  • Source: ITT Visual Information Solutions

– Idrisi Taiga

  • Version: 16.0
  • Source: Clark University

– Global Mapper

  • Version: 12.0
  • Source: American Outland

– Surfer: Version: 9.0
– MIKE model set. Source: Danish Institute of Hydraulics

(b) **Build the flood hazard zoning map for the whole province**
    Flood risks zoning is based on an analysis of each region, each unit-surface, and
    the combined impact of the following factors:

– Topography, shape of the basin, flow direction – shown in **the contour map of
  slope classification**.
– Soil permeability, ability of flow generating, erosion, and washing-off are shown
  in **the classification map of soil permeability**.
– Buffer surface, vegetation and human impact factors (deforestation, slash-and-
  burn, mining, structural construction, houses, bridges, etc.) shown in **the map of
  vegetation cover**.
– From the rain data of the monitoring stations in Dak Lak province, make a map
  of rain isometric then classify the rain value according to the level affecting the
  flash flood generation through **the rainfall classification map** (Fig. 4.1).

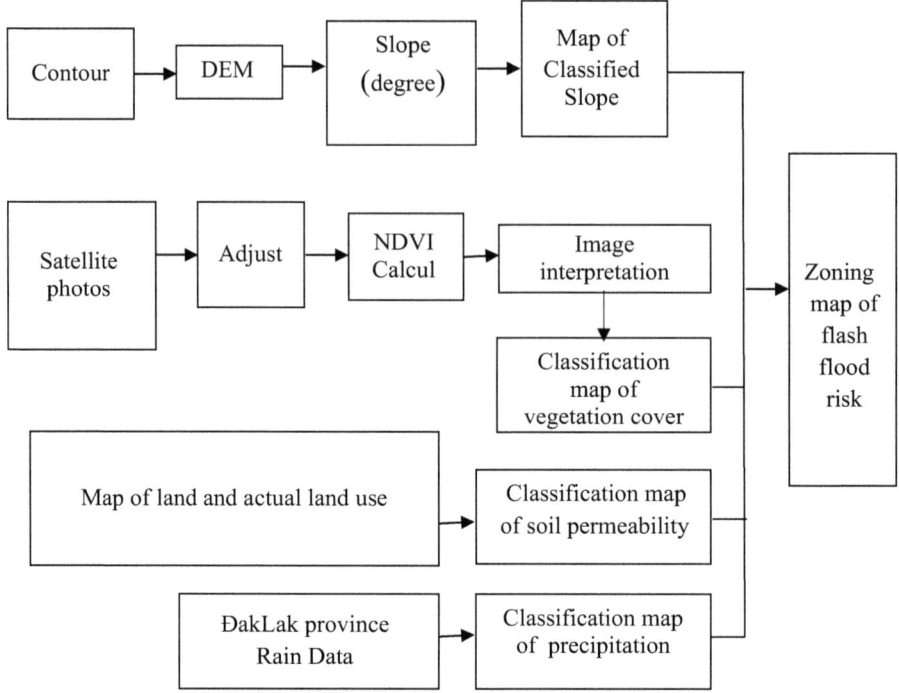

Fig. 4.1 Steps to create the zoning map of flash flood risk

4.1.3 Build a Digital Elevation Model

(a) **Concept of the digital elevation model**

Digital elevation model (DEM) or digital topography model (DTM) is a digital model used to describe the topographical elements, including elevation and slope (gradient and slope direction). DEM is one of the important applications in geographic information system (GIS), that is, the ability to model the topographic surface based on the capacity to organize data and methods in GIS. In the establishment steps of the flash flood risk zoning map, the construction of the digital elevation model DEM is in the second step.

(b) **Sense of a DEM building in flash flood research**

In this study, the GIS technique is applied to build the digital elevation model of the basin –such as a basis for calculating the topographical parameters of the basin, as well as to calculate other parameters of the basin. Therein, the most important is to calculate the slope of the basin, then combine with other parameters such as vegetation cover, rock, and soil to help predict the risk of flash floods of some areas in the research region.

(c) **Steps in a DEM building**

Study uses the method of DEM building from the contour map. Using the XYZ Extractor software transforms the vector data layer .TAB into the file XYZ As same as click 'toado' ('coordinate' in Vietnamese). XYZ has the coordinates (x, y, z) of all vertexes, nodes of the contours and elevation points. Then convert file.XYZ to file.txt (Fig. 4.2).

Surfer 9.0 and Global Mapper 12.0 software was used to interpolate irregularly discrete data of the zonal space into a regular network and place the interpolation data into files with the extension .GRD. Grid file is used to create contour lines and surfaces (Figs. 4.3, 4.4, and Table 4.1).

Comment on Elevation in Dak Lak Province

The topography of Dak Lak has the following main forms:

• **Highland topography**

Considered the most typical topography of Dak Lak, it can be divided into the following terrain styles:

– Terrain at an altitude of 100–300 m, accounting for 26.17%, mainly includes areas such as Ea Soup and some areas along the Vietnam-Cambodia border.
– Terrain at an altitude of 300–500 m, accounting for 32.56% in the basin with Lak valley.
– Terrain at an altitude of 500–800 m high, accounting for 30%, including Buôn Ma Thuot city and Pleiku plateau, which is one of the two largest plateaus in the Central Highlands covered by a basalt layer with a fairly flat, inclined surface.

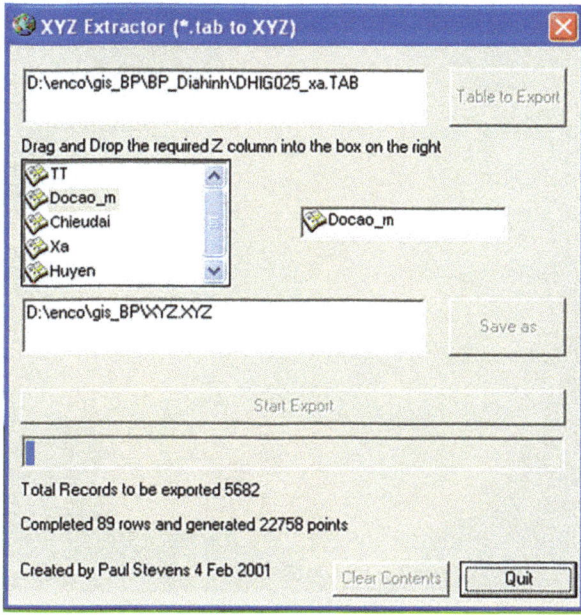

Fig. 4.2 Program XYZ Extractor gets coordinates of the vector file nodes in the linear form

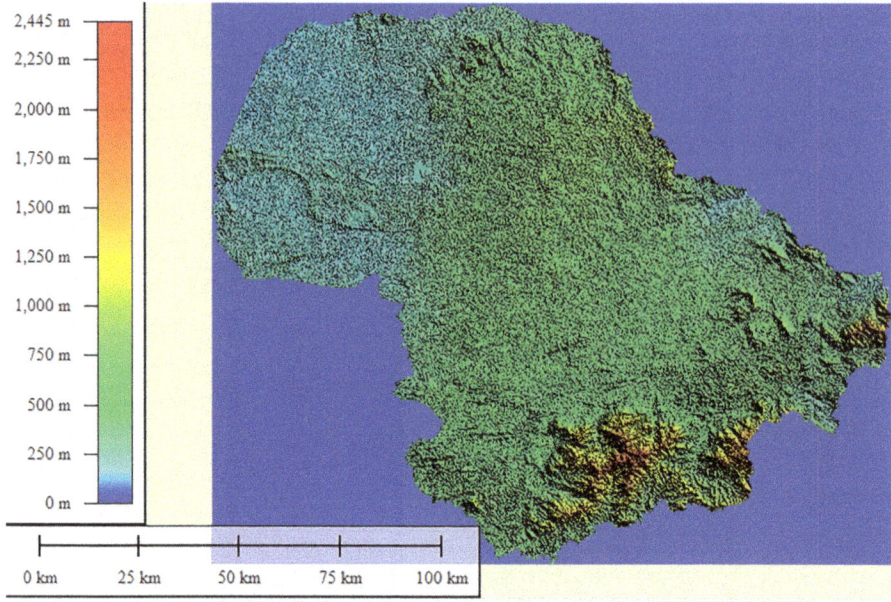

Fig. 4.3 Model of elevation in Dak Lak province from top view

Fig. 4.4 The photo shows the digital elevation model in Dak Lak province

Table 4.1 Tables and maps show percentage (%) of the altitude in Dak Lak province

Code	Altitude (m)	Pixel quantity	Surface (km²)	Ratio (%)
1	0–300	8,649,208	3447.062	12.22
2	300–500	10,761,973	4289.086	46.50
3	500–800	10,019,577	3993.210	35.24
4	800–2445	3,622,382	1443.667	5.96

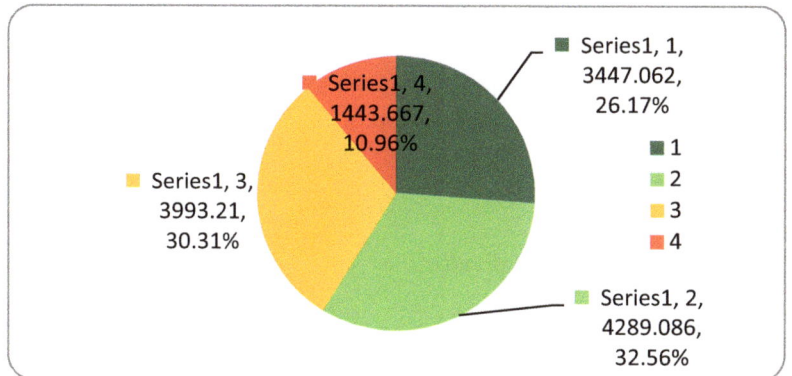

Gradually the southward has an altitude of 400–600 m, while the north and northeast are 750–800 m. Buôn Ma Thuôt Plateau is also a large basalt plateau that runs from north to south over 90 km, from east to west about 70 km.

- **Mountainous terrain**

The region has an altitude of 800–2445 m, accounting for 10.96% of Dak Lak surface, and there are some high mountains separating the basin. The Western Khanh Hoa mountains has a Ca Dung mount peak of 1978 m, Gia Lô peak of 1817 m. The West Khanh Hoa mountain chain is a drainage divide of water basin between the Krông Ana river and Da Nhim river. The Chư Yang Sin mountain chain is formed by a granite block on the south of Krông Pak-Lăk low-land area extending in the direction of northeast-southwest and has the mountains with the highest peak of 2405 m.

Chư Yang Sin Mountains have a vertical steep on the northern Krông Ana valley, and gradually lower to the west, and to the south of Chư Yang Sin chain is Krông No valley. Dan Sona-Ta Dung Mountains have a vertical steep on the northern Krông Kno valley, and gradually lower to Lang Biang and Dji Ring plateaus.

In general, Dak Lak's topography is strongly divided; the stream network is thick, originating from many small branches focusing on the big branch. These are very favorable condition for flash floods in this area.

4.1.4   Build a Slope Map

(a) **Data processing**

The slope of basin terrain is important to the drainage process. If the basin has a steep slope, when it rains, water along the slopes drains quickly into the main river, whereas on a small slope, water drains more slowly.

In the algorithm, the changes of terrain elevation in two directions x, y are the parameters to determine the mountain side direction and the terrain slope at a point. Here consider the value of Z height as a function of two coordinates (x, y), which can be expressed as $Z = f(x, y)$.

The slope map is calculated on the basis of the DEM of the basin elevation model, which calculates the terrain slope on the original image using the spatial analysis tool Gis Analysis/Surface Analysis/Topographic Variable/Slope of Idrisi Taiga software (Fig. 4.5).

(b) **Form a map**

From the slope map, the area is divided into four slope grades (Fig. 4.6 and Table 4.2).

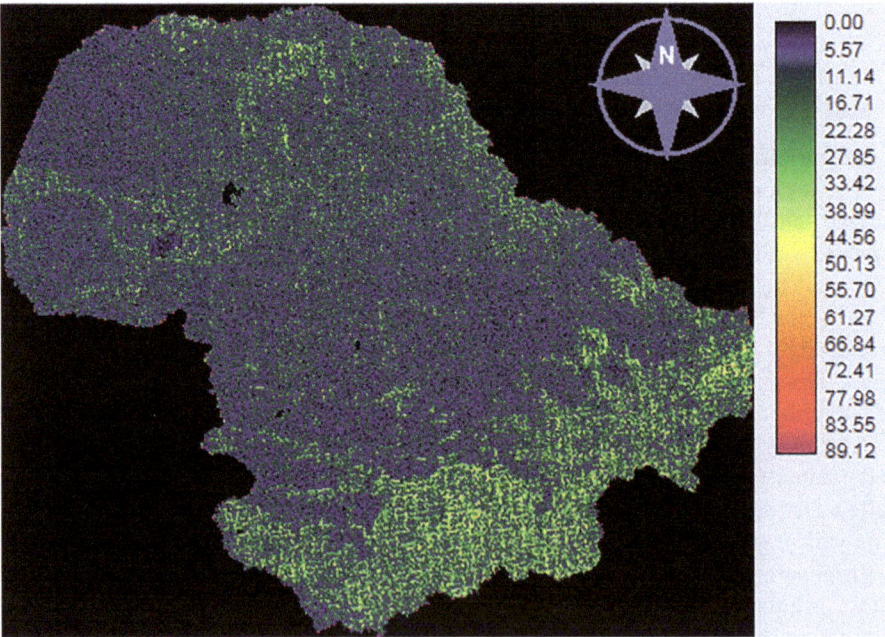

Fig. 4.5 The photo shows the terrain slope in Dak Lak province

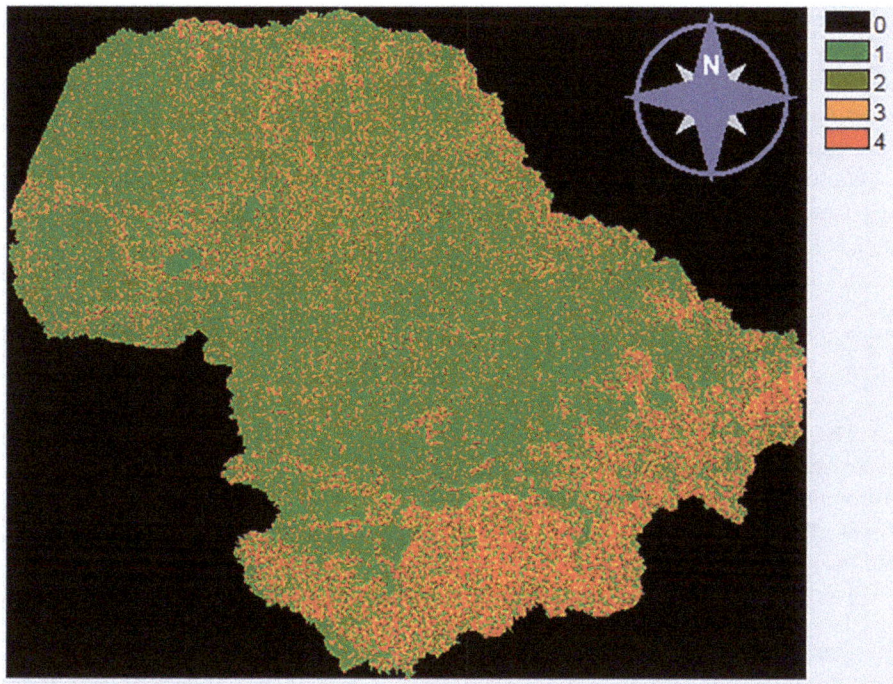

Fig. 4.6 Image of slope level distribution at M'Drăk district

Table 4.2 Table and graphs showing the proportions of slope level values in Dak Lak province

Impact level	Low	Medium	High	Very high
Slope value (degree)	0–7	7–15	15–25	25–90
Quantity of pixels	10,560,721	13,198,036	6,086,475	3,050,866
Surface (km²)	4208.878	5259.956	2425.709	1215.897
Ratio %	32.10	40.12	18.50	9.27
Code	1	2	3	4

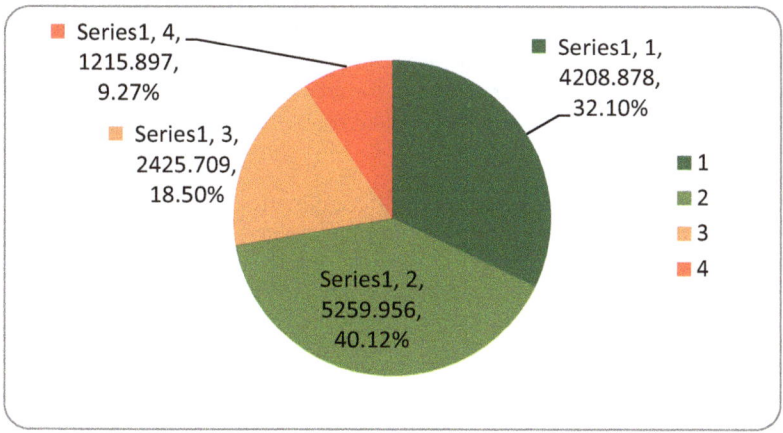

Comment

Looking at this slope classification map we see that the Dak Lak topography has a very steep slope. The mountainous terrain is concentrated mainly in the southeast, focusing on the districts of Lak, Krông Bông, and M'Drăk, this is also where the Phuong Hoang Col is bordering Khanh Hoa province.

We see that the high level (18.50%) and very high level (9.27%) in this grading map occupy a quite large area, so we see the terrain here can very easily lead to erosion, soil washing-off when it rains and gives great speed to surface runoff.

4.1.5 Build a Vegetation Cover Map

(a) The concept of vegetation index – NDVI

To evaluate whether the vegetation cover is high or low, the Normalized Difference Vegetation Index (NDVI) is used. NDVI is determined based on the different reflection of plants expressed between the visible spectrum (usually the red) and the sub-infrared spectrum.

Plant index NDVI is calculated using the following formula:

$$NDVI = (Near-infrared\,channel - red\,channel)/(Near-infrared\,channel + Red\,channel)$$

SPOT 4 images include:

- The near infrared range is band 3
- The red range is band 2

$$So: \boxed{NDVI = (Band\,3 - Band\,2)/(Band\,3 + Band\,2)}$$

NDVI is in the range $[-1,1]$.

NDVI has a negative value indicating that the visible range has a higher reflectivity than that of the near-infrared range, where there is no vegetation (Fig. 4.7).

A value of $0 < NDVI < 1$ indicates where the near-infrared range has a higher reflectivity than the visible range, indicating that the area has vegetative cover. The closer the NDVI value to 1, the thicker the area of vegetation (Fig. 4.8).

In summary, only positive values are suitable for areas with vegetation. Negative values are due to clouds, water, snow, arid soil, and rocks.

(b) Building the image interpretation key

The interpretation key is built on natural color combinative images.

Fig. 4.7 Negative NDVI values on 2011 Spot4 image

Water: On the natural color combinative image, the water is blue, the river has a linear shape and smooth structure (Fig. 4.9).

Clouds: On the natural color combinative image, the clouds are white, with speckled shapes and black clouds shadow (Fig. 4.10).

Bare land without vegetation: On the natural color combinative image, bare land is white in color, with indeterminate shape (Fig. 4.11).

Construction land: On the image of combinative natural color, construction land is white in color, with indeterminate shape (Fig. 4.12).

Vegetation: On the natural color combinative image, vegetationis green. Depending on the green level shown, we can classify the regions of very thick, thick, and medium vegetation distribution (Figs. 4.13, 4.14, 4.15, and 4.16).

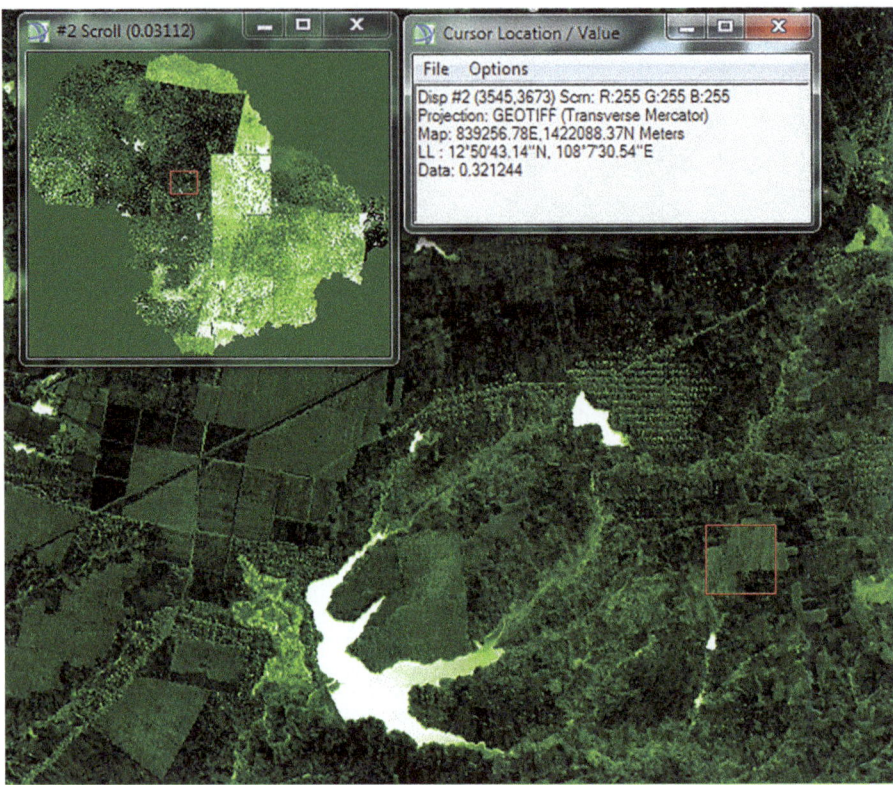

Fig. 4.8 Positive NDVI value on Spot4 image in 2011

As stated in the previous section on the natural conditions of the Dak Lak region, this area does not have large water surfaces, so small water surfaces with NDVI ≤ 0 value will be counted in the NDVI value level with high influence on flash flood formation. We will use image interpretation elements (shadows) to identify the clouds, and then assign its NDVI value to the same main object surrounding it (Fig. 4.17).

(c) **Building the hierarchical map of vegetation cover**

A hierarchical map of vegetation cover is built by using the satellite images (Spot 4) to calculate the NDVI vegetation index and interpreting the image classification. Then decentralizing objects according to the influence level for the flash flood formation (Table 4.3).

In the table above, as smaller vegetative cover factor (NDVI decreasing) corresponds to the poor water holding capacity and erosion resistance, the greater the impact of this factor on the flash flood formation.

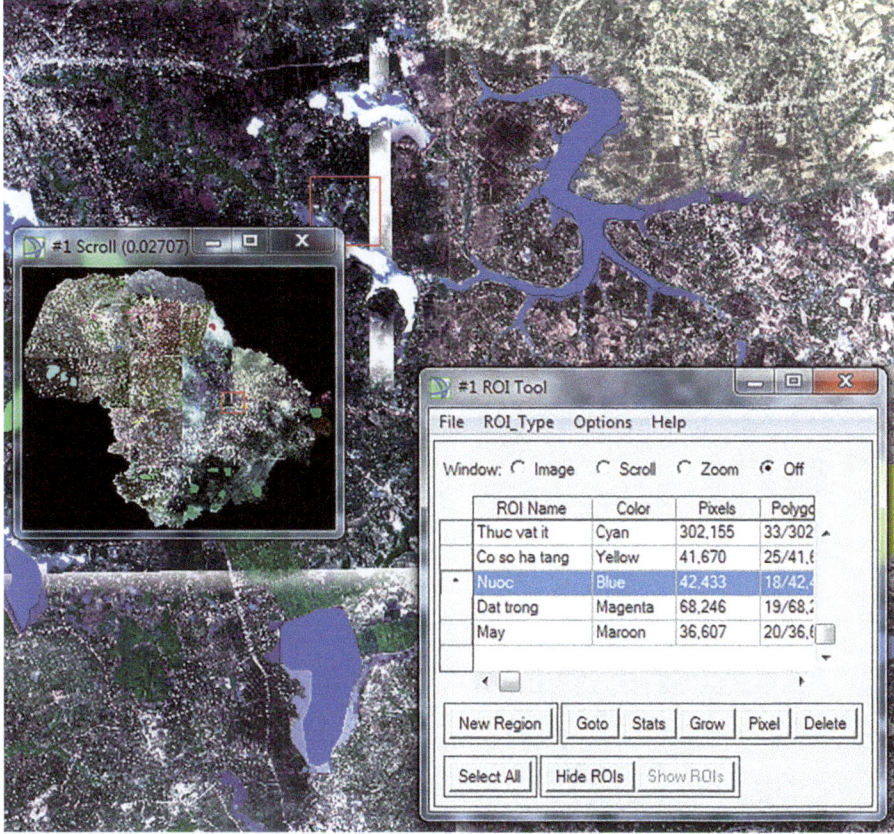

Fig. 4.9 Illustration of water object interpretation key

After vegetative cover is graded, the image is detached according to the boundary of the study area. The image data model is Raster, so it will be kept to serve map overlapping. We can divide the regional vegetation map into four grades (Fig. 4.18 and Table 4.4):

Comment

We found that the vegetation coverage of Dak Lak province is quite poor, with the lowest coverage locations in the district in the west and northwest. The two grades of NDVI value (code 3, code 4) have the greatest influence on the formation of flash floods that occupy a large area. Therefore, when it rains, the amount of rainwater retained by plants is negligible and creates conditions that easily cause ground erosion, contributing to the creation of solid matter for flood flows.

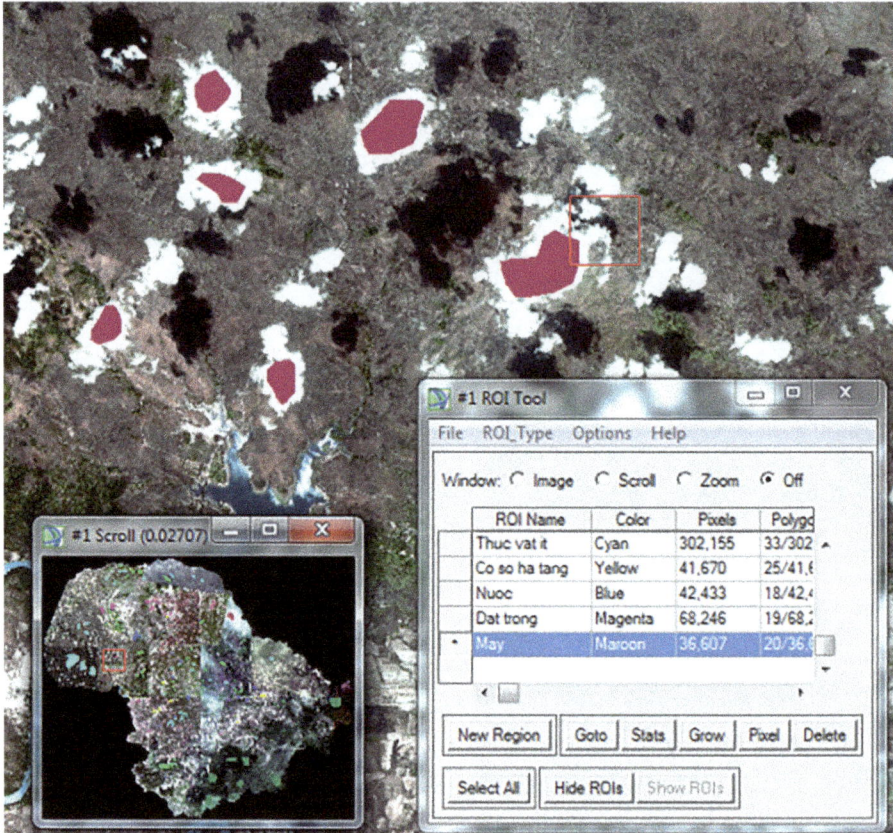

Fig. 4.10 Illustration of cloud object interpretation key

4.1.6 Building the Classification Map of Soil Permeability Speed

The classification map of soil permeability speed was developed from the soil map and land use map of the study area. Soil permeability speed is decided mainly on the mechanical composition of the aforesaid soil. Therefore, from the soil types available in the study area, we will determine its mechanical composition, thereby deducing the permeability speed for each soil type (according to the classification of scientist Nguyen Trong Yem).

When considering the soil permeability speed in an area, pay attention to factors such as structures on the land (houses, industrial parks, factories, irrigation works with water surface, etc.) or natural surface covers of the land (natural ponds, lakes, rivers, and streams) because they have large surfaces that prevent contact between rainwater and soil.

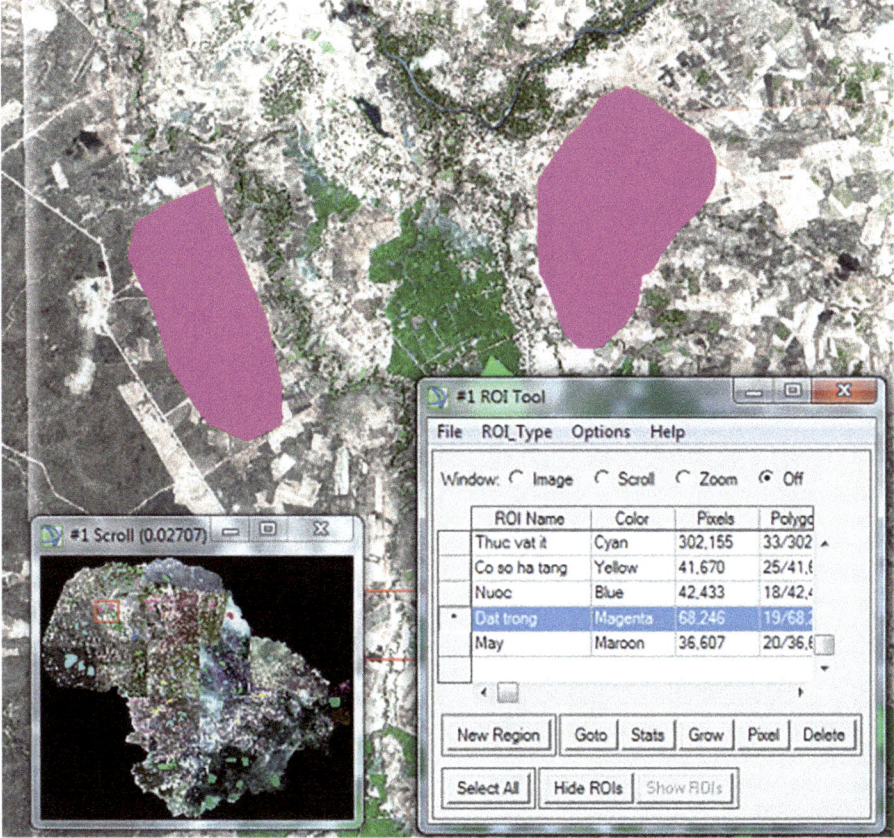

Fig. 4.11   Illustration of a key that explains a bare land object without vegetation

If they have a large and concentrated land cover area, we will remove these areas because they can hardly create large currents, so in the calculation their value will be assigned to 0. As for the areas with these structures occupying a small and sparse area, we can ignore their presence because it does not affect the general seepage rate of the soil on that area.

After having the permeability speed values of the soil types in the study area, we build data for the soil permeability speed layer by assigning the obtained seepage speed values on the soil map using Mapinfo and convert data from Vector model to Raster and classify the seepage rate values into groups by using Idrisi Taiga software (Table 4.5).

According to the table above, the smaller the soil permeability speed, the greater the creative ability of surface runoff, so the greater the impact on flash flood formation (Fig. 4.19 and Table 4.6).

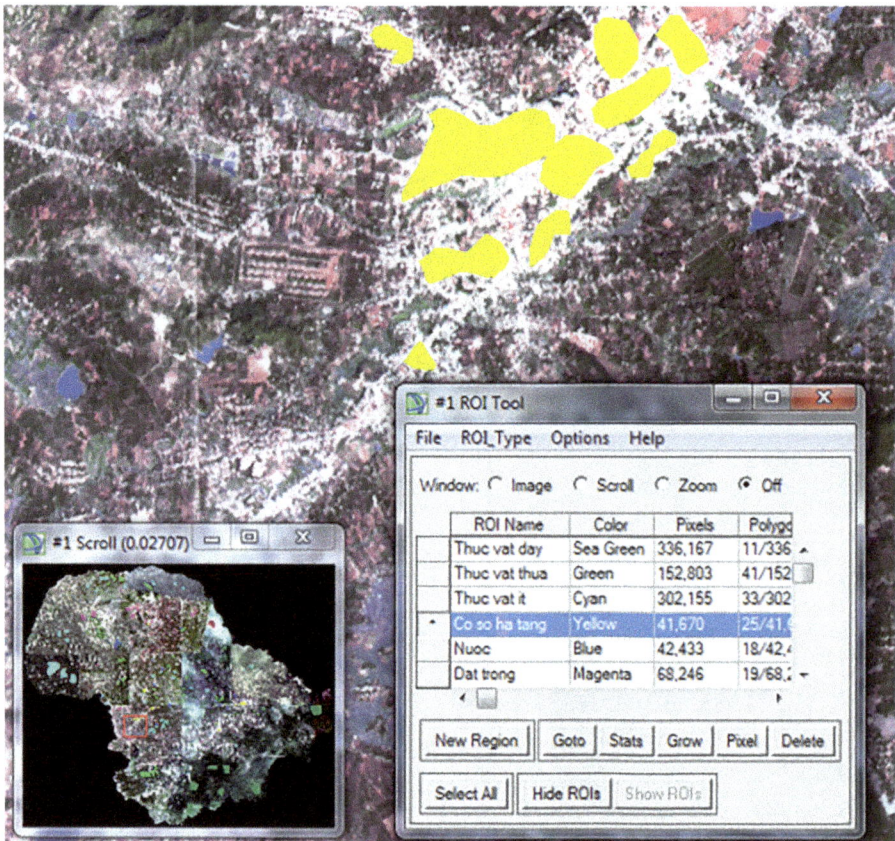

Fig. 4.12 Illustration of construction land object with interpretation key

Comment

From the classification results and the distributive images of the permeability speed value, we see that the number of surfaces with a low level of influence (code 1, code 2) on the flash flood formation is mainly occupied. However, the ratio of the soil type surfaces having a high level of influence on flash flood formation is not small (code 3, code 4) and mainly concentrated in the districts of M'Đrăk, Krông Năng, Ea. H'Leo, and Krông Ana.

4.1.7   Build a Classification Map of Average Rainfall

Rainfall classification maps are often built based on two data sources: hourly and daily rainfall data corresponding to two types of rain classification that cause flash floods, which is the classification according to the largest period of rain (rain

Fig. 4.13 Illustrates the interpretation key of very thick vegetation objects

threshold) or according to the biggest rain day. For the hourly rain data, we will use statistics for 1, 3, 6, 12, and 24 hours as follows (Table 4.7):

If the rainfall reaches the above correspondent thresholds, rain will likely cause flash floods. But in this rainfall classification, we can only divide two levels: YES and NO capable to cause flash floods, so in order to match the evaluation of the factor roles in the topic, we will use the classification method. Rainfall according to rainy days (Table 4.8).

In the table above, the rain effect on the flash flood formation is increasing with the value of rainfall. In order to use reliable daily rain data, according to hydrological experts, the rainfall data must be at least measured over a period of 15 years.

From the daily rainfall data at eight pluviometer stations of Dak Lak province and on-place survey, the statistics of rains <200 mm, < 350 mm, < 450 mm, and ≥ 450 mm and the highest rainfall threshold has the number of rainy days reaching the frequency of 1%, the respective impact level will be selected for calculation. After obtaining the rainfall value of the study area, we build rain layer data (zonal form) using Mapinfo software and convert data from Vector model to Raster using

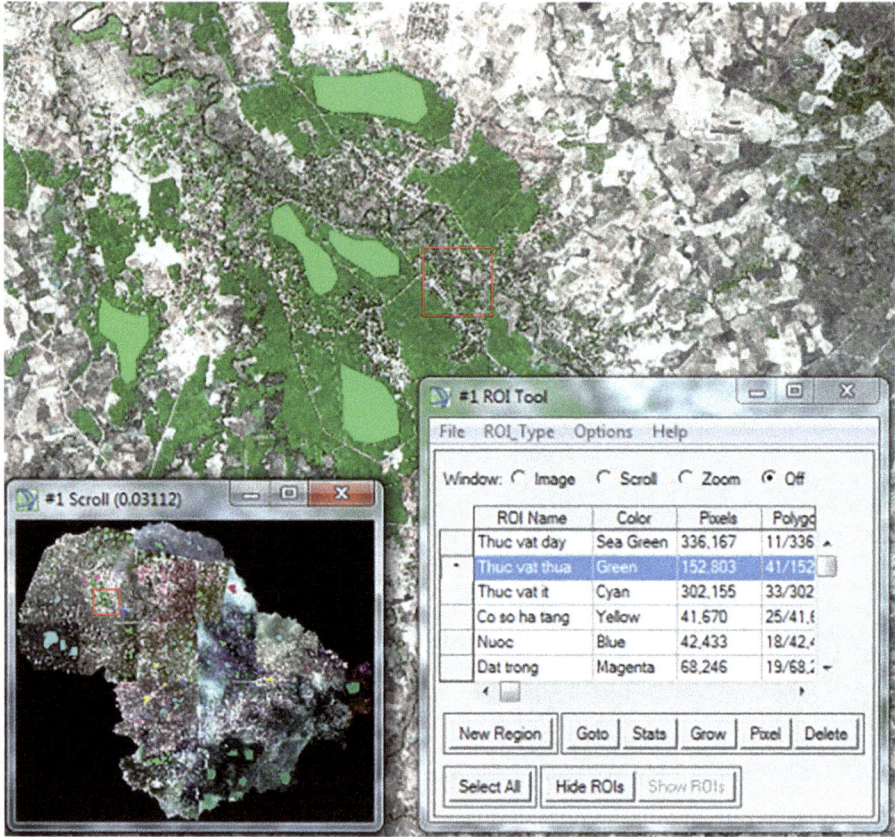

Fig. 4.14 Illustration of interpreting key of thick vegetation object

modules of Reformat/RasterVector of Idrisi Taiga software for Raster overlapping work (Fig. 4.20).

Through the daily rainfall data of heavy rains in Dak Lak province, the main rainfall is less than 450 mm, so when classifying the rainfall value we only have three codes (code 1, 2, and 3) (Table 4.9).

Comment

We see that in Dak Lak, although the average daily rainfall is low (92.58%), we still see the sudden occurrence of rainy days with high intensity, such as: M'Drăc district 430 mm (November 25, 2008), Buôn Hô 280.5 mm (August 5, 2007), Lăk 264.5 mm (October 10, 2000), and Ea H'Leo 205 mm (September 29, 2009). Other favorable conditions can cause flash floods.

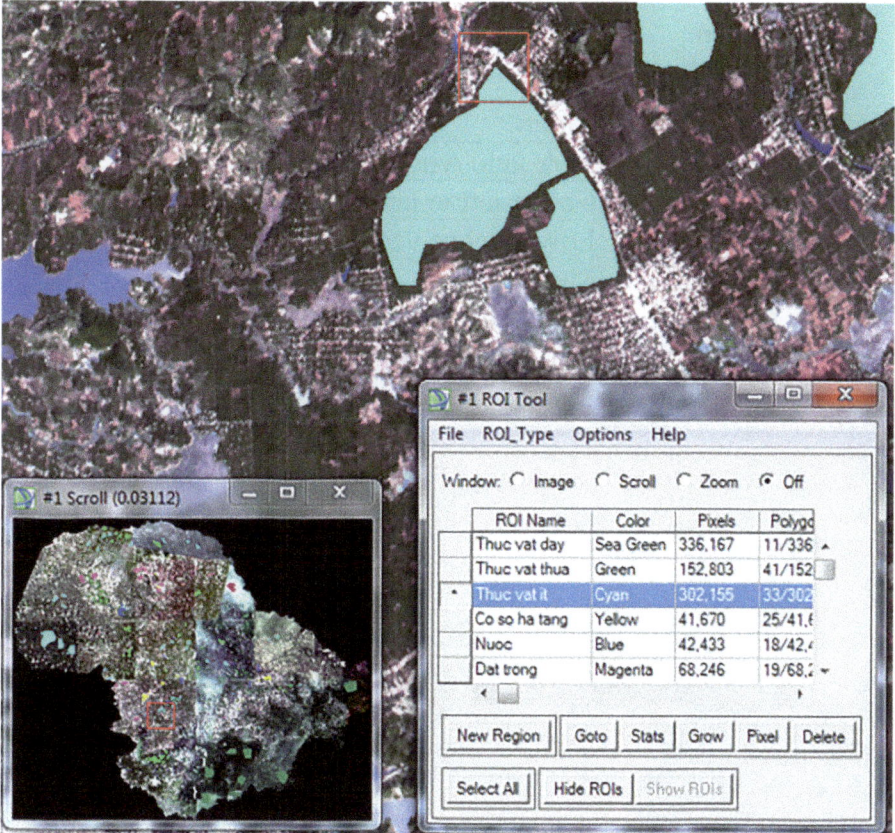

Fig. 4.15 Illustration of the interpretation key for medium vegetation object

4.2   Build the Dak Lak General Zoning map of Flash Flood Risk

– **Zoning basis of flash flood risk**

The risk zoning map is built on the basis of a 4 layer overlapping map of elemental factors that cause flash floods:

– Slope classification map.
– Vegetation cover classification map.
– Map of rainfall by day.
– Soil permeability classification map.

– **Steps of realization**

In order of aggregative information and map formation of flash flood risk, we convert all maps into Raster form with the pixels quantity, row, and column coordinates. Then use the arithmetic calculation to add the element maps together.

Color	Notes
	Water
	Cloud
	Bare land without vegetation
	Construction land
	Low level of vegetation
	Medium level of vegetation
	Thick vegetation

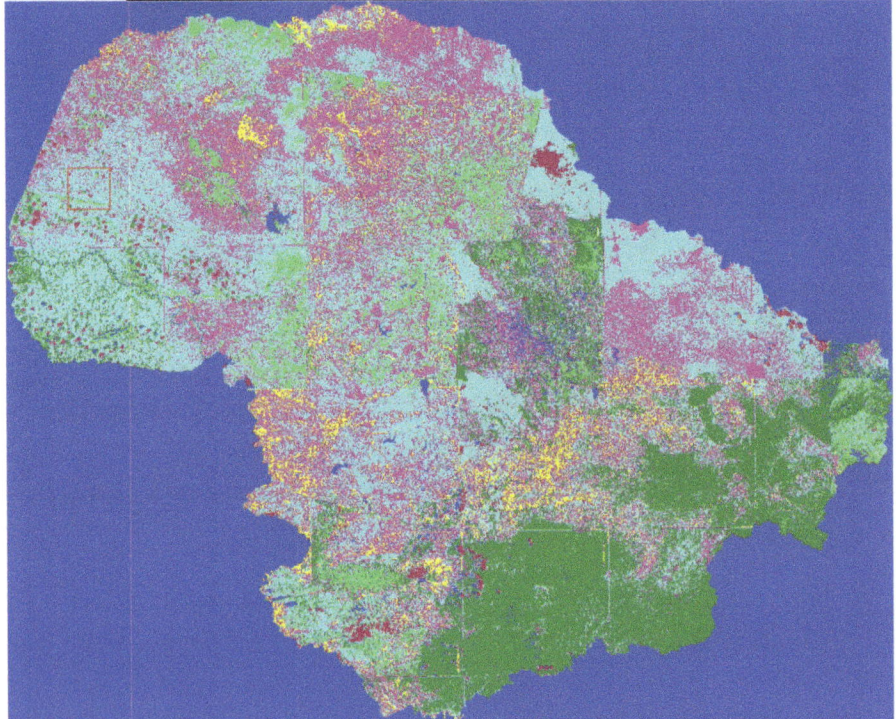

Fig. 4.16 Interpretation results of Spot 4 images in 2011

Determining the weights for each map: According to field research results in seven districts of Dak Lak province where flash floods occurred, the factor of vegetation cover and rainfall plays a key role in forming flash floods. With the experience of the research team doing two similar topics in Quang Ngai and Binh Phuoc provinces, the weights are determined as follows:

Flood map = vegetation cover classification map-value (1.5) + map of soil permeability value classification + slope value classification map + rain value classification map value (1.5). Each map has different levels of classification:

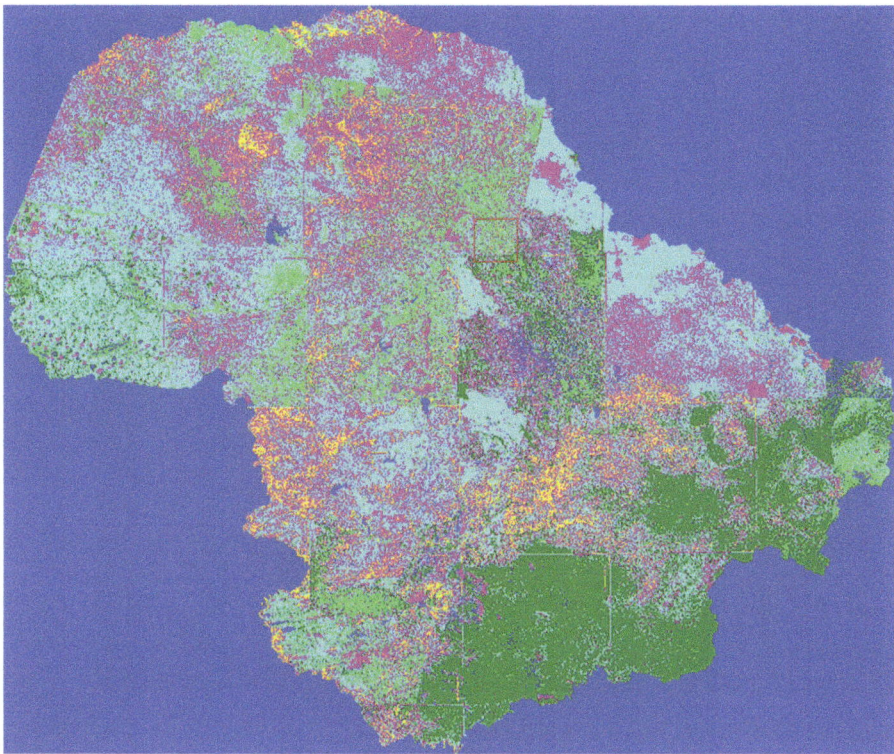

Fig. 4.17 Spot 4 image interpretation results have eliminated clouds in 2011

Table 4.3 NDVI value classification in accordance with influence level for flash flood formation

No.	NDVI values	Objective type	Impact level	Code
1	0.8 to 1 (Add NDVI value of cloud cover surface – if any)	Old forest, regenerating forest in many layers	Low	1
2	0.5 to 0.8 (Add NDVI value of cloud cover surface – if any)	Perennial gardens, planted forests, etc.	Medium	2
3	0.2 to 0.49 (Add NDVI value of cloud cover surface – if any)	Garden land around the house, short time trees after harvest	High	3
4	−1 to 0.2 (Subtract NDVI value of cloud cover surface – if any)	Bare land, urban areas, small water surface, etc.	Very high	4

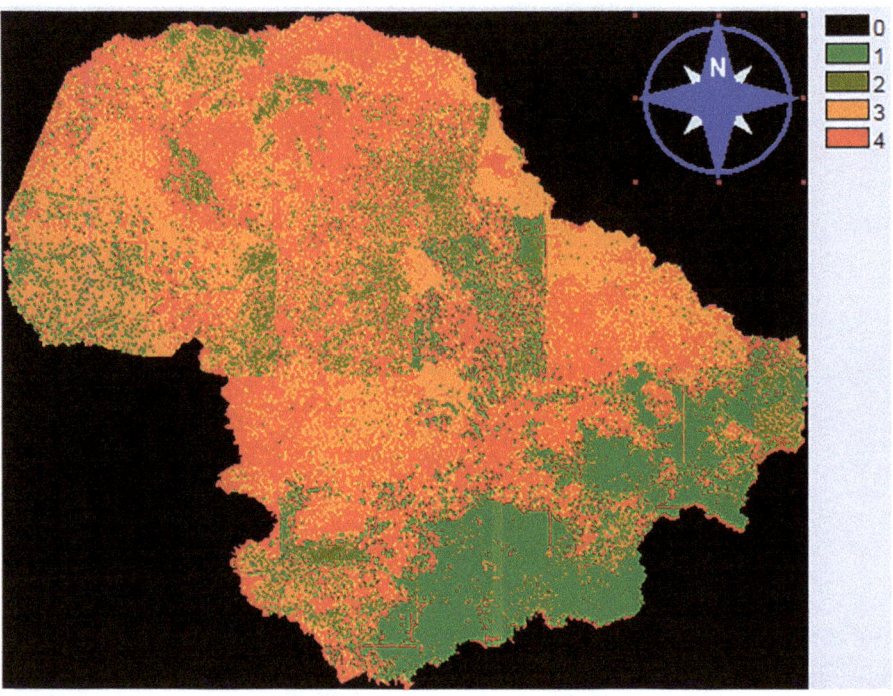

Fig. 4.18 Vegetation classification map in Dak Lak province

Table 4.4 Table and graph of ratios of vegetation classification values in Dak Lak province

Impact level	Low	Medium	High	Very high
Pixel quantity	6,077,015	2,759,645	12,715,680	11,343,758
Surface (km²)	2421.939	1099.831	5067.717	4520.951
Ratio %	18.47	8.39	38.65	34.48
Code	1	2	3	4

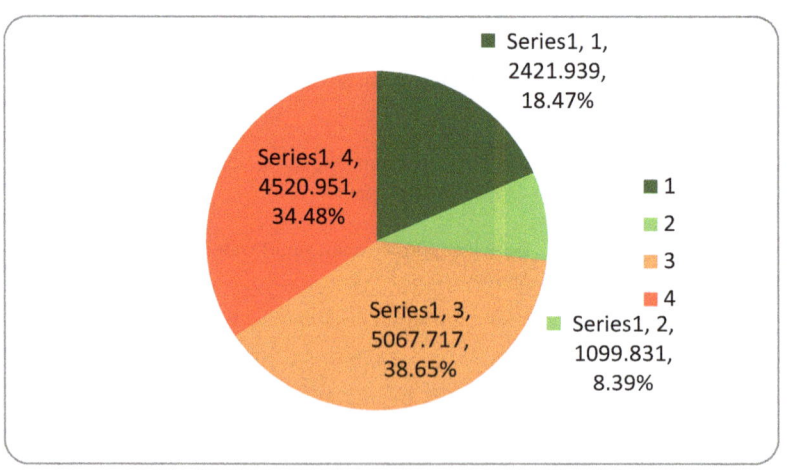

Table 4.5 Table of classification of soil seepage speed in the province of Dak Lak

No.	Type of land	Permeability speed	Impact level	Code
1	Gray soil and infertile gray soil	Mush seeped	Low	1
2	Alluvial soil, brown soil semi-arid region	Medium seeped	Medium	2
3	Black soil, red soil (red-brown soil on basalt), soil group with tight clay layer, heterogeneously mechanical	Little seeped	High	3
4	Eroded rocky soil, glay soil	Very little seeped	Very high	4

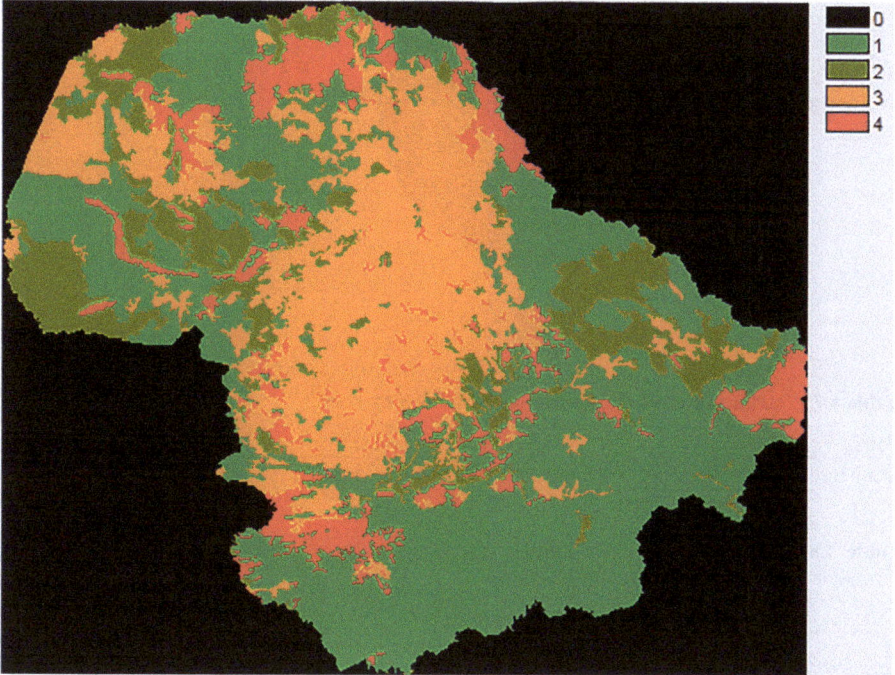

Fig. 4.19 Map of distribution of permeability speed level in Dak Lak province

The slope map has four levels:

- Level 1: 00–70.
- Level 2: 70–150.
- Level 3: 150–250.
- Level 4: > 250.

Vegetation cover map has four levels:

- Level 1: area with very thick vegetative cover.
- Level 2: the area with thick vegetation cover.
- Level 3: medium vegetation cover.
- Level 4: bare land with no vegetation, area covered by clouds, and thin vegetation.

Table 4.6 Table and chart of ratios of permeability speed values in Dak Lak province

Impact level	Low	Medium	High	Very high
Pixel number	14,880,691	5,135,047	9,465,488	3,414,872
Surface (km²)	5930.563	2046.526	3772.383	1360.966
Ratio %	45.24	15.61	28.77	10.38
Code	1	2	3	4

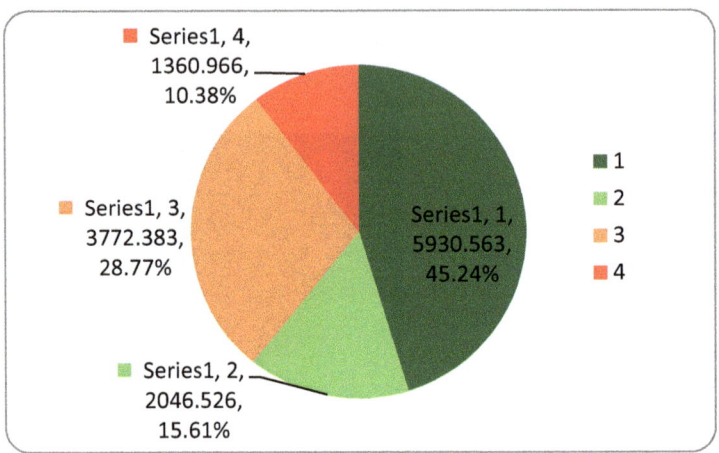

Table 4.7 The threshold of rain that causes flash flood by hour

Hour	1	3	6	12	24
Rain threshold (mm)	100	120	140	180	220

Table 4.8 Classification of daily rainfall according to the influence level for the flash flood formation (according to Nguyen Trong Yem)

No.	Daily rainfall value (mm) with frequency $P = 1\%$	Influence level	Code
1	<200	Low	1
2	200–350	Medium	2
3	350–450	High	3
4	≥ 450	Very high	4

Rainfall distribution map haves three-level:

- Class 1: < 200 mm.
- Level 2: 200–350 mm.
- Level 3: 350–450 mm.

Land classification map has four levels:

- Level 1: gray soil and degraded soil.
- Level 2: alluvial soil, reddish brown soil in semi-arid regions.
- Level 3: black soil, red soil, heterogeneous mechanical tight clay layer.
- Level 4: eroded soil that is inert, gravel, or glay.

Fig. 4.20 Map of rainfall level distribution in Dak Lak province

Table 4.9 Table and graph of the rate of rainfall supply values in Dak Lak province

Influence level	Low	Medium	High	Very high
Pixel quantity	30,455,417	2,337,694	102,623	0
Surface (km²)	12137.726	931.666	40.899	0
Ratio %	92.58	7.11	0.31	0
Code	1	2	3	4

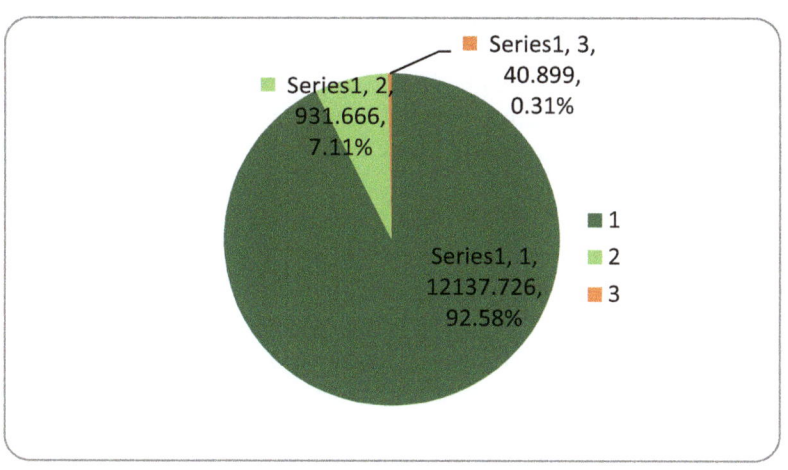

Table 4.10 Classification table of flash flood risk level

Code	Risk level	Value range
1	Low	1–5
2	Medium	5–10
3	High	10–15
4	Very high	15–18.5

Table 4.11 Tables and graphs of flash flood risk values in Dak Lak province

Risk level	Low	Medium	High	Very high
Pixel quantity	1,642,387	13,398,071	15,949,429	2,022,650
Surface (km²)	654.5583	5339.678	6356.498	806.1086
Ratio %	4.98	40.58	48.31	6.13
Code	1	2	3	4

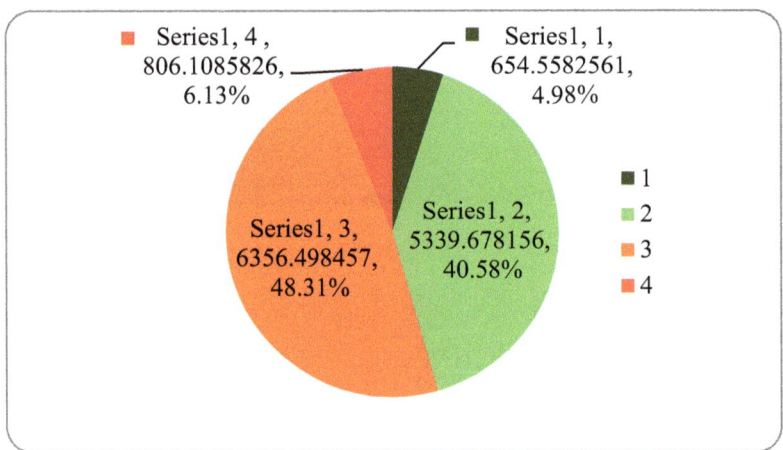

Level 4 of the slope hierarchy map has the highest risk of flash floods in basins with a slope of over 30% (27°), with cover less than 10% on level 4 of the vegetation cover zoning map, and soil permeability is very low at level 4 of the zoning map of soil types. There is prolonged rain with a total volume of more than 3000 mm, including the rainfall rate per day greater than 200 mm.

After map overlapping, we have a result image with a value from 0 to 18.5 (maximum 4 * 1.5 + 4 + 4 + 3 * 1.5).

The current classification is as follows:

+ Level 1: 1–5 (Maximum 1 * 1.5 + 1 + 1 + 1 * 1.5)
+ Level 2: 5–10 (Max 2 * 1.5 + 2 + 2 + 2 * 1.5)
+ Level 3: 10–15 (Max 3 * 1.5 + 3 + 3 + 3 * 1.5)
+ Level 4: 15–18.5 (Table 4.10, 4.11, Figs. 4.21 and 4.22).

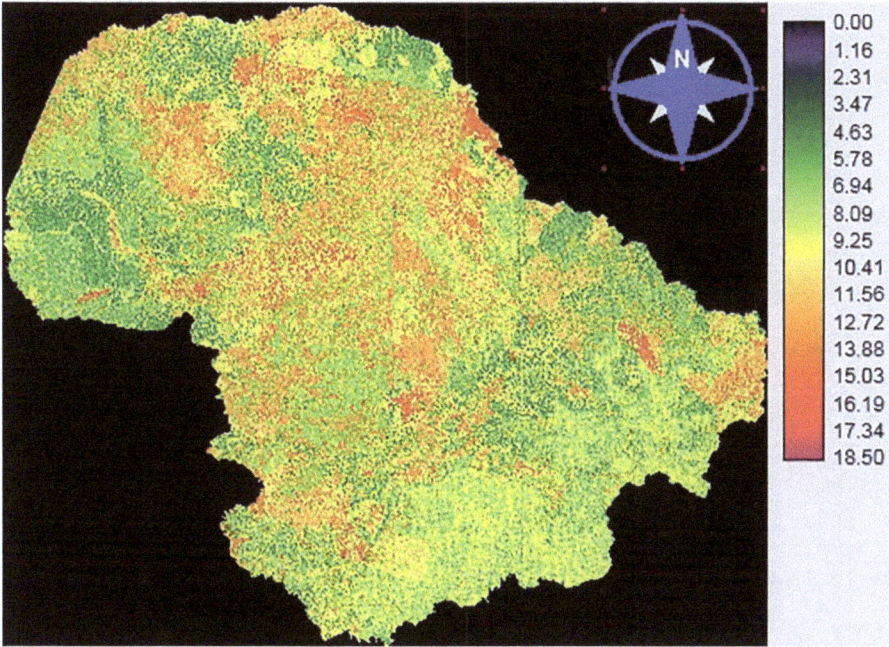

Fig. 4.21   Image of the distribution of flash flood risk in Dak Lak province

Fig. 4.22   Distribution of value level of flash flood risk in Dak Lak province

Table 4.12 Table of areas at risk levels for each district in Dak Lak province

No.	Name of district	Code 1 surface (km^2)	Code 2 surface (km^2)	Code 3 surface (km^2)	Code 4 surface (km^2)	Total surface (km^2)
1	Cư M'gar	7.723 (0.94%)	202.121 (24.53%)	492.731 (59.81%)	121.295 (14.72%)	823.87
2	Ea H'Leo	56.872 (4.25%)	334.033 (24.97%)	829.031 (61.98%)	117.657 (8.80%)	1337.59
3	Krông Năng	5.517 (0.91%)	157.672 (25.91%)	369.694 (60.75%)	75.667 (12.45%)	608.553
4	M'Đrắk	66.700 (5.00%)	651.964 (48.83%)	563.180 (42.18%)	53.201 (3.98%)	1335.05
5	Lắk	42.353 (3.37%)	606.300 (48.30%)	554.659 (44.18%)	52.074 (4.15%)	1255.438
6	Ea Súp	112.226 (6.36%)	652.962 (37.03%)	946.159 (53.66%)	51.936 (2.96%)	1763.28
7	Krông Pắc	27.288 (4.35%)	225.573 (35.95%)	325.763 (31.92%)	48.770 (7.77%)	627.394
8	Buôn Đôn	165.273 (11.75%)	759.027 (53.97%)	434.897 (30.92%)	47.160 (3.35%)	1406.36
9	Buôn Ma Thuột	7.912 (2.10%)	140.684 (37.30%)	195.329 (51.80%)	33.194 (8.80%)	377.119
10	Krông Buk	0.624 (0.18%)	88.120 (25.73%)	222.542 (64.99%)	31.161 (9.10%)	342.447
11	Krông Bông	35.152 (2.80%)	703.368 (55.97%)	488.275 (38.85%)	29.939 (2.38%)	1256.73
12	Krông Ana	14.570 (4.09%)	116.613 (32.77%)	197.738 (55.57%)	26.916 (7.76%)	355.837
13	Ea Kar	91.823 (8.86%)	439.030 (42.37%)	486.187 (46.92%)	19.176 (1.85%)	1036.22
14	Buôn Hồ	0.088 (0.03%)	94.635 (31.99%)	182.841 (61.81%)	18.264 (6.17%)	295.828
15	Cư Kuin	9.984 (3.46%)	140.391 (48.64%)	123.882 (42.92%)	14.363 (4.98%)	288.62

Comment

The risk ratio of flash floods is very high, accounting for 6.13%, mainly distributed in the districts of Cư Mgar, Krông Năng, Lak, M'Drăc, and Ea H'Leo. This region has a steep slope, limited vegetation cover, and concentrated rainfall. The risk of flash floods is high, accounting for 48.31%, scattered in the province. The average risk area accounts for 40.58% and the low risk area accounts for 4.98% in Buôn Dôn and several parts of the province (Table 4.12).

Comment

The flash flood risk map in Dak Lak province shows that the capacity of a flash flood in this area is quite high. Which can be divided into the following three groups:

– Districts at very high risk (with a large area of grade 4) are Cư M'gar (14.72%), Ea H'Leo (8.80%), Krông Năng (12.45%), M 'Dak Lak (3.98%), and Lăk (4.98%).

- The districts with potential flash floods are Ea Sup, Krông Pac, Buôn Dôn, Buôn Ma Thuôt, and Krông Buk.
- The districts that are less likely to have flash floods are Krông Bông, Krông Ana, Ea Kar, Buôn Hô, and Cư Kuin.

  Compared with the field survey data, this result is completely correct.

4.3   Build the Risk Zoning Map for Key Points Area

Based on the field survey data and using the GIS method, we can see that the two areas most vulnerable to flash floods and damage in recent years are Cư Mgar district, which has the Srepôk river system, and the Krông Năng district, which has a system of rivers Krông Năng and Ba passing through.

4.3.1   Flash Flood Zoning Map of Cư M'gar District

The flash flood zoning map of Cư Mgar district is divided into four grades (Fig. 4.23 and Table 4.13):

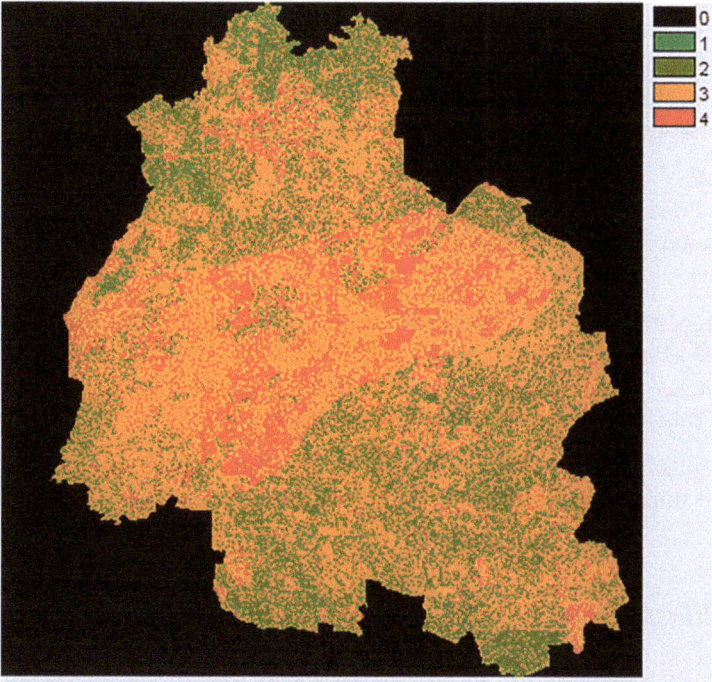

Fig. 4.23 Value distribution of value level of flash flood risk in Cư Mgar district

Table 4.13 Table and graph of value ratio of flash flood risk in Cư Mgar district

Risk level	Low	Medium	High	Very high
Pixel quantity	19,379	507,154	1,236,340	304,348
Surface (km²)	7.723	202.121	492.731	121.295
Ratio %	0.94	24.53	59.81	14.72
Code	1	2	3	4

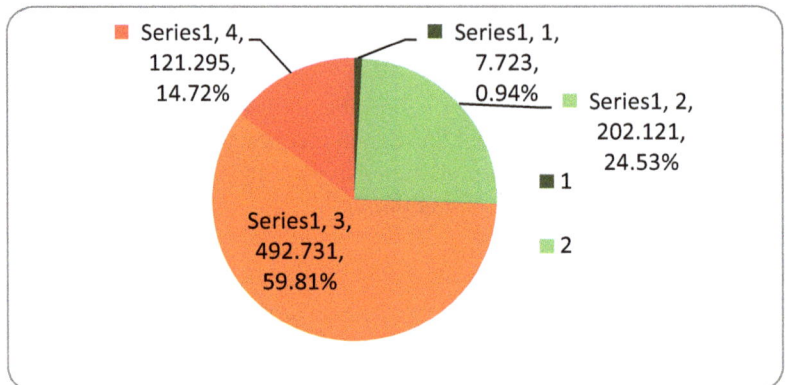

Comment

The risk ratio of flash floods is very high, accounting for 14.72%, mainly distributed in communes with steep slopes, limited vegetation, and concentrated rainfall. The risk of flash floods is high, accounting for 59.81%, scattered in the province. The medium risk area is 24.53% and the low risk area is 0.94% (Table 4.14).

Comment

The flash flood risk map and actual data collected in Cư M'gar district shows that the likelihood of flash flooding in this area is quite high, which can be divided into two groups:

– Communes at high risk (large area of level 4) are Ea H'dinh, Ea Tar, Quang Hiep, Cư Dliê M'nong, Ea Kiet, Cư M'gar, Ea KPam, Ea M'DRóh, and Ea M'năng.
– Communes with potential flash floods are Ea Kuêh, Ea D'Rông, Ea Tul, Cư Sue, Cuôr Đăng, Quảng Tiến, Ea Pok, and Quảng Phú.

The results from the map are completely consistent with the survey results on the flash flood situation in Cư Mgar district, Dak Lak province.

4.3.2 Flash Flood Zoning Map in Krông Năng District

The Krông Năng district flash flood zoning map is divided into four grades (Fig. 4.24 and Table 4.15):

Table 4.14 Table of areas at risk levels for each commune in Cu Mgar district

No.	Name of commune	Code 1 surface (km²)	Code 2 surface (km²)	Code 3 surface (km²)	Code 4 surface (km²)	Total surface (km²)
1	Ea H'đinh	0 (0%)	1.164 (2.71%)	26.578 (61.92%)	15.181 (35.37%)	42.923
2	Ea tar	0 (0%)	2.625 (6.41%)	23.167 (56.58%)	15.150 (37.00%)	40.942
3	Quảng Hiệp	0.304 (0.53%)	4.607 (8.02%)	37.744 (65.72%)	14.776 (25.73%)	57.431
4	Cư Dliê M'nông	0 (0%)	7.845 (12.82%)	41.469 (67.78)	11.872 (19.40%)	61.186
5	Ea Kiết	2.498 (2.75%)	24.510 (26.99%)	52.713 (58.04%)	11.101 (12.22%)	90.882
6	Cư M'gar	0 (0%)	5.983 (19.25%)	16.231 (52.23%)	8.863 (28.52%)	31.077
7	Ea Kpam	0 (0%)	7.198 (17.06%)	25.777 (63.01%)	7.932 (19.39%)	40.907
8	Ea M'DRóh	0 (0%)	8.963 (16.55%)	37.793 (69.78%)	7.405 (13.67%)	54.161
9	Ea M'nang	0.033 (0.15%)	4.530 (19.92%)	12.803 (56.29%)	5.377 (23.64%)	22.743
10	Ea Kuêh	4.500 (4.03%)	33.217 (29.77%)	69.471 (62.26%)	4.398 (3.94%)	111.586
11	Ea D'Rơng	0 (0%)	25.204 (36.24%)	40.063 (57.61%)	4.274 (6.15%)	69.541
12	Ea Tul	0.003 (0.01%)	17.187 (30.26%)	36.627 (64.48%)	2.989 (5.26%)	56.806
13	Cư Suê	0.030 (0.09%)	17.555 (50.21%)	15.159 (43.35%)	2.222 (6.35%)	34.966
14	Cuôr Đăng	0 (0%)	14.441 (44.19%)	16.207 (50.21%)	2.033 (6.22%)	32.681
15	Quảng Tiến	0.005 (0.02%)	9.060 (35.30%)	15.124 (58.92%)	1.480 (5.77%)	25.669
16	Ea Pốk	0.347 (0.85%)	14.235 (34.98%)	25.258 (62.08%)	0.849 (2.09%)	40.689
17	Quảng Phú	0 (0%)	3.789 (39.01%)	5.539 (57.03%)	0.385 (3.96%)	9.713

Comment

Krông Năng district has a very high risk of flash flooding. The ratio of flash flood is very high, accounting for 18.92%, mainly distributed in the central and northeastern communes of the district. The risk of flash floods is high, accounting for 64.10%, scattered in the province. The medium risk area accounts for 18.92% and the low risk area accounts for 3.26% (Table 4.16).

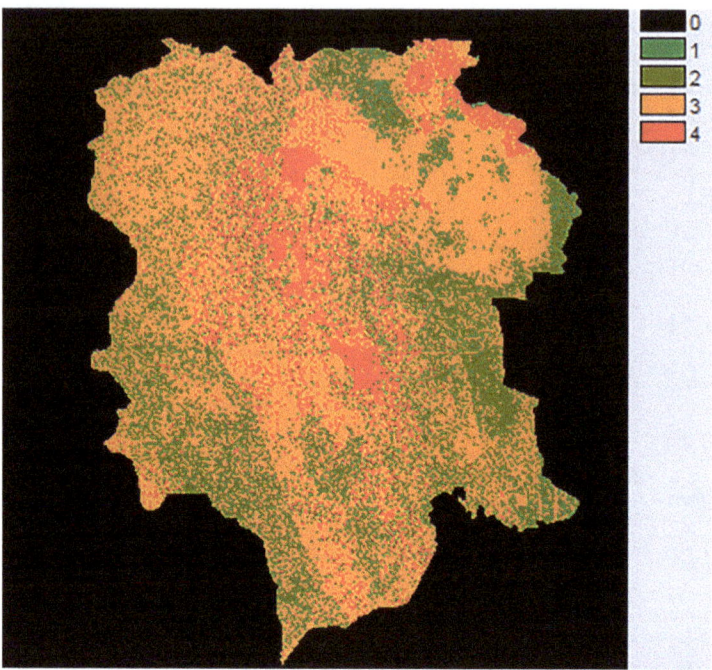

Fig. 4.24 Value distribution according to the risk level of flash flood in Krông Năng district

Table 4.15 Table and graph of flash flood risk values in Krong Nang district

Risk level	Low	Medium	High	Very high
Pixel quantity	13,845	395,625	927,621	189,862
Surface (km²)	5.517	157.672	369.694	75.667
Ratio %	0.91	25.91	60.75	12.43
Code	1	2	3	4

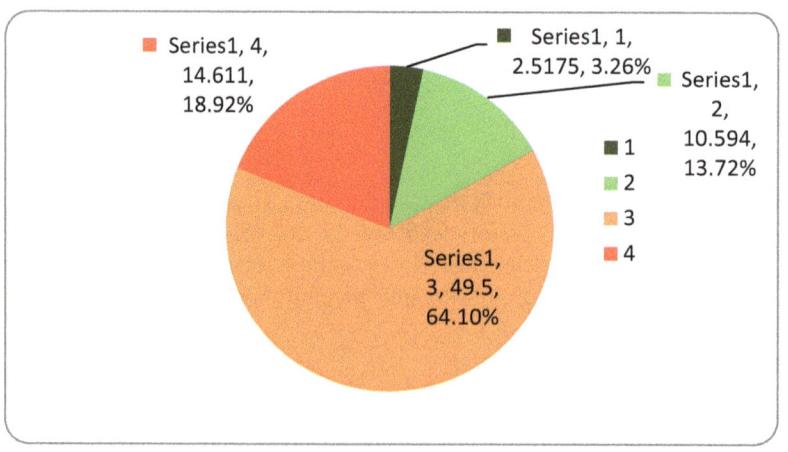

Table 4.16 Table of areas at risk levels for each commune in Krong Nang district

No.	Name of commune	Code 1 surface (km²)	Code 2 surface (km²)	Code 3 surface (km²)	Code 4 surface (km²)	Total surface (km²)
1	Dliê ya	0.866 (1.08%)	11.615 (14.53%)	47.946 (59.97%)	19.528 (24.42%)	79.955
2	Cư Klông	2.5175 (3.26%)	10.594 (13.72%)	49.500 (64.10%)	14.611 (18.92%)	77.222
3	Ea tam	0.949 (1.14%)	24.567 (29.39%)	46.674 (55.84%)	11.388 (13.63%)	61.186
4	Tam Giang	0 (0%)	3.611 (10.58%)	22.364 (65.53%)	8.155 (23.89%)	34.13
6	Ea Puk	0.123 (0.28%)	17.729 (40.72%)	19.420 (44.61%)	6.262 (14.38%)	40.907
7	Phú Lộc	0 (0%)	3.206 (9.67%)	24.958 (75.28%)	4.991 (15.05%)	33.155
8	Phú Xuân	0.084 (0.19%)	22.368 (49.80%)	18.698 (41.63%)	3.762 (8.38%)	44.912
9	Ea Dáh	0.916 (1.77%)	16.196 (31.30%)	31.004 (59.92%)	3.623 (7.00%)	51.739
10	TT Krông Năng	0.059 (0.24%)	12.333 (49.77%)	9.051 (36.52%)	3.338 (13.47%)	24.781
11	Ea Tân	0 (0%)	9.551 (17.65%)	42.520 (78.58%)	2.039 (3.77%)	54.11
12	Ea Tóh	0 (0%)	6.664 (16.84%)	30.053 (75.92%)	2.867 (7.24%)	39.584
13	Ea Hồ	0 (0%)	19.232 (45.98%)	21.502 (51.40%)	1.097 (2.62%)	41.831

Comment

The likelihood of a flash flood in Krông Năng district is quite high, which can be divided into the following two groups:

- Communes at relatively high risk (with large area of grade 4) are Cư Klông, Dli ya, Ea Tam, Tam Giang, Ea Puk, Phu Loc, Phu Xuan, Ea Dăh, and Krông Năng Town.
- Communes likely to have flash floods are Ea Tóh, Ea Tân, and Ea Hô.

The results from the map are completely consistent with the survey results on the flash flood situation in Krông Năng district of Dak Lak province. Areas that need special attention to these techniques are the high-risk areas where many ethnic minorities live and produce agriculture.

- The basin has the streams Ea Tul and Ea Huar. Starting from 700 m upstream (Cư Dlie Mnông commune) flowing down 200 m downstream (Ea Huar commune), the flash flood area includes communes: Cư Dlie Mnông, Ea Tar, Ea Kpam, Ea Hđinh, Ea Kiêt, Cư M'Gar, Ea Mroh, Quang Hiep, Ea M'năng (belonging to Cư M'Gar district). Ea Huar, Ea Wer (Buôn Dôn District).

– Krông Năng river basin. Flowing from 750–800 m in the upstream, to 400–500 m downstream, through the communes: Dliê Ya, Cư Klông, Ea Tam, Phu Loc, Tam Giang, Ea Puk, Krông Năng Town, Phu Xuan, Ea Dah belongs to Krông Năng and Xuan Phu districts, Ea Kar town of Ea Kar district.
– Krông Jing Stream originates from the altitude of 500 m in Cư M'tar, Cư Kroa commune, falling to a height of 400 m in Krông Jing commune and M'Drăk town.
– Stretching along Dăk Phơi stream, Dăk Liêng in Lak district. Upstream with the altitude of 1000–1200 m (Dăk Phơi commune, Bông Krang) down to 400–500 m altitude in the two communes of Dak Nuê and Dak Liêng and a part of Dak Liêng town.

Note

There are many different solutions to prevent and limit the harmful impacts of flash, tubing floods. However, it must be based on specific natural, social conditions and flood causes in each locality, plus a flexible coordination among these solutions. For example, for places with high slope characteristics, solutions include to segment flood flows, separate solid objects from flood flows, build control dams, and plant forests. For areas with bare land and hills, land reclamation and afforestation must be carried-out. For areas prone to flash floods near residential areas, carry out evacuation, migration, and resettlement in safer areas or build structures and houses with waterproof walls.

4.4 Building a Data Based on Potential Risk Capacity

Map System

– Land map.
– Map of land use.
– Map of water system.
– Administrative map of Dak Lak province.
– Map of the actual soil cover in Dak Lak province.
– Map of slopes classification in Dak Lak province.
– Map of average rainfall per year of the province.
– Map of soil permeability classification.
– Zoning map of flash flood risk in Dak Lak province.
– Zoning map of flash flood risk in Cư Mgar district.
– Zoning map of flash flood risk in Krông Năng district.

Data System of Rain, Soil, and Hydrology

Use specialized software for GIS and RS: Mapinfo 9.0, Arcgis 10.0, Envi 7.0, Idrisi Taiga 16.0, Surfer 9.0, and Global Mapper 12.0.

Chapter 5
Build the Map of Flash Flood Risk (Case Study: Dak Lak Province)

Abstract This chapter presents a method of mapping flash flood risk based on existing data (of Chap. 4), based on the requirements, principles, and presentation of a current status and forecast map. The strengths of remote sensing techniques (RS), aero-images interpretation, geographic information systems (GIS), digital models, specialized software, etc., combined with survey results and actual data analysis are used to make a map of the flash flood risk.

The chapter lists 10 contents of map-building from processing input data sources, synthesizing analysis, and simulation for the representations on flash flood risk maps, so as to have an image basis for formulating a prevention framework strategy, for damage minimization, and provide the observation capacity into potential risks. The chapter also outlines the approach methods and applicable techniques to present a typical flash flood map, for example of Dak Lak province, at a scale of 1/10,000 for the whole province and 1/25,000 for high-risk areas. In addition to a map there are recommendations for technical options to prevent flash floods.

Keywords Map building · Forecast map · Applicability of GIS, RS · Terrestrial database exploration

5.1 Requirements of a Flash Flood Risk Map

Terrestrial database exploration and collection on time of devastating flash flood phenomenon are difficult challenges. This has promoted the application of aerial observation and spatial remote sensing techniques for monitoring and mapping the areas at flash flood risk. The satellite imagery to inventory flood areas is very important work when collecting enough real data about events will affect establishing and validating the flash flood susceptibility model. Historical information on flash flood occurrences is the basis of the validation of flood sensibility models.

The inventory map can be made from different sources: field data collection, historical storage, local community information, and satellite image interpretation. In recent years, satellite imagery has increasingly been used in the analysis of

© The Author(s), under exclusive license to Springer Nature Switzerland AG 2022
L. H. Ba et al., *Flash Floods in Vietnam*,
https://doi.org/10.1007/978-3-031-10532-6_5

hazardous areas. With the development of the spatial resolution of satellite imagery with gradually higher resolution, different types of problems can be analyzed, investigated, identified, and extracted, which helps to adjust and verify the susceptibility of the built model. The high-resolution remote sensing data in different types can help in the preparation of the inventory map for different types of natural disaster risks.

Every current study uses two methods for the flood inventory map. The first one uses historical data sources plus field investigation, regional community information, reports and data collected by functional department, combined with the interpretation of high-resolution satellite images. The second one includes a detailed assessment of the study area that analyzes an individual case of historic over-peak flood, using high-resolution satellite images with the help of field investigation.

5.1.1 Contents of Forecast Map of Flash Flood

A base map is a corresponding topographic map that can simplify some unnecessary elements:

- Shows all types of formed flash floods.
- Shows the intensity of flash floods.
- Shows the formative probability of flash flood.

Thus, the shown content must have three basic types of flash floods: Congestive flow flash floods, slope side flash floods, and complex flash floods. Complex and congestive flash floods occur only at the appropriate locations and with a factor of congestion, so indicated at those locations. Slope side flash flood is assessed on the basis of a combination of influencing factors that should be presented on the surface. The main factor to evaluate flash floods is the flash flood intensity coefficient or a combination of factors.

5.1.2 Principles of Flash Flood Mapping

The flash flood map is made:

- Based on the nature of flash flood formation and development;
- Based on the assessment of a combination of affecting factors of flash floods formation and development.

The studies show that there are three basic types of flash floods: congestive flash floods, slope side flash floods, and complex flash floods. The detailed classifications of other authors can all be combined into these three basic types of flash floods. Because the types of flash floods have a completely different nature of formation and development, the association principle shown on the zoning map is also different.

Thus, the combination of factors influencing flash flood formation and development is summarized as follows:

– Necessary conditions for flash floods: water sources, such as rain, melted snow, and broken reservoirs water.
– Sufficient condition is a buffer surface factor, including topography, vegetation cover, and weathering-soil crust. The geological-tectonic factors are the scientific basis for determining the above-mentioned buffer elements.

These two conditions must have binding compatibility to form and develop flash floods in terms of type, intensity, and probability.

The scale of the map is based on the purpose and area of the study area: With regional forecast for mountainous regions, the map scale can be about 1/250,000–1/500,000. For a large region or river basin, it could be on a scale of 1/50,000–1/100,000. For small area or small river basin, it can be more detailed such as a scale of 1/25,000.

5.1.3   Description on Forecast Map of Flash Flood

It is very important for the map to show clarity and transparency and help the reader clearly see the content that needs to be shown. Regardless of topography, flash floods are represented by the following factors:

– The points with different magnitudes show the intensity of flash floods, the color of points shows the danger level for people and asset;
– The color tones and marked brush represent the distribution area of flash floods;
– Tables and diagrams;
– Notes.

5.2   Profit by Ability and Applicability of GIS, RS

The geographic information system (GIS) creates a faster and cheaper map on the basis of the previously used map. Where there is no professional drawing staff, it is still possible to do so with the help of a computer. With thematic maps, using the superposition method, a map takes more general purpose and contains more information. For example, when there are layers describing rivers, traffic, and land resources of a province stacked together, you get a complete map of a province.

• Convenient to make and update maps when the data is in digital form.
• Allows to represent different graph types in the same data.

- Convenient for data analysis that requires interaction between statistical analyses with maps.
- Storing and displaying information completely separate. Information can be displayed at different scales.
- The most important advantage of GIS is that its data is always current and the combinative ability of data sources effectively.
- GIS is used in the following fields: business, precinct zoning, infrastructure management, map and database publishing, resource management and urban, environmental planning.

- **The role of GIS in flash flood research**

The most basic function of GIS is to store, process, and present data in the most general way. All data from various sources such as field surveys and satellite imagery (remote sensing) can be integrated and stored in the GIS database in the form of digitized Vector and Raster layers, from maps, satellite images (zonal, linear), enter information of survey points (key-point objects).

Based on the researcher's knowledge or the software data analysis before putting into the geographic information system, GIS analyzes and presents data such as integration, zoning of objects or object groups classified according to their calculation, the quality of each layer of data and according to research purposes.

- One of the important applications of GIS techniques in flash flood research is to build a digital elevation model. DEM is built on the structure of Raster images (rows, columns), which can be exactly reconstructed with a fairly accurate terrain compared to reality, which is the basis of constructing derivative maps such as slope maps and side direction.
- Construction of DEM in Dak Lak province combined with some other GIS applications such as flow density mapping, topographical division, and quantitative calculation of basin parameters (coefficient of form, concentration, and bend). Moreover, DEM also contributes to development planning in Dak Lak province, as well as the environmental forecasts such as flood due to rain, erosion, accretion, and landslides.
- The advantage of GIS is that it can permanently update data and overlay data in order to most generally assess the correlation among the factors that cause flash floods such as topography, geology, and vegetation cover. Therefore, we can evaluate the combined effects of all these factors.
- GIS creates a final product by zoning the places where flash floods are possible from small to large levels and stores data of each risk level based on calculation analysis, and assesses the correlation of all factors that cause flash floods, using the regional features. From there, it is possible to evaluate the distribution and surface of each level and establish the thematic maps to analyze the current state of flash floods in the basin.
- From the zoning map of flash flood risk, we can combine with many other maps such as real land use map to serve the schematic planning of that area and propose the appropriate environmental protection measures.

– Remote sensing application (RS): using remote sensing images.

– In summary, the most basic and important application of GIS in flash flood research is to build and manage databases of all factors related to flash floods, typically topography and overlay distribution of vegetation and geology and to synthesize the elements of flash flood classification map.

5.3 Contents, Access Method and Techniques of GIS, RS

We have completed the Flash Flood Risk Mapping study for three provinces: Binh Phuoc, Quang Ngai, and Dak Lak. Here only the results of Dak Lak province are introduced as one case study.

5.3.1 Proposed Contents

– **Content 1**: Collecting, investigating, and studying the natural features and socio-economic conditions related to flash floods in Dak Lak.
– **Content 2**: Surveying, investigating, and analyzing the characteristics of key areas, artificial reservoirs, hydro-electricity, and irrigation systems on rivers and fountains in the province to identify the risks of provincial flash floods.
– **Content 3**: Assessing the current situation, analyzing developments of the factors that mainly affect the flash flood formation in Dak Lak province.
– **Content 4**: Research and assessment of the real flash flood status in the entire Dak Lak province.
– **Content 5**: Analyzing the flash flood potential is based on closely related factors.
– **Content 6**: Flash flood simulation of some typical flash floods occurred in some main river basins.
– **Content 7**: Research to establish a forecast map of the potential flash floods arisen in Dak Lak province, scale 1/100,000.
– **Content 8**: Building the flash flood risk maps for key areas and two main river basins (for example, with Dak Lak being the Srepok and Ba River sub-basins, with Binh Phuoc being Bu Dop and with Quang Ngai being five to six districts on West Mountains) with the scale 1/25,000.
– **Content 9**: Developing an overall framework strategy for solutions to prevent and reduce damage caused by flash floods in the province (Dak Lak, Quang Ngai, and Binh Phuoc), period 2011–2020.
– **Content 10**: Building flash flood potential database in Dak Lak province.

5.3.2 Approach, Method, and Technique Used

(a). **Content 1**: Collecting, investigating, and studying the natural features and socio-economic conditions related to flash floods in Dak Lak.

Approach Manner

To complete the project to build a zoning map of flash flood risk in a certain locality, the necessary conditions are basic data, data on natural and socio-economic conditions. Using data that can preliminarily assess the situation of flash floods in the province, effective prevention plans can be built. The data should satisfy these criteria:

– Ensure objectivity, reliability, sufficient length, continuity, representative of the research area.
– Particularly, data on forms that cause heavy rain such as storms and tropical depressions need to be collected on a large scale, ensuring spatial consistency.

Method

Data collection was carried-out over a period of 14 months (from December 2010 to February 2012), mainly statistics and comparison.

Techniques

The work of statistics and comparison was mainly assigned to the clerical staff to perform. The data on topography, land, and vegetation are listed in different application forms. Each form type has its own statistics, ensuring scientific integrity and easy to recognize data characteristics.

(b). **Content 2**: Surveying, investigating, and analyzing characteristics of key areas, artificial water reservoirs, hydro-electricity, and irrigation systems on rivers and fountains in the province to identify the flash flood risks in the area.

Approach, method, and implementation technique:

Surveying, investigating the flood traces, and identifying areas prone to flash floods are very important to build zoning maps of flash flood risk in an area. It determines the precision of the map, the land area that was destroyed by flash floods, its material mass carried away.

Therefore, the approach manner should clearly state that the investigation purpose is to identify key areas where flash floods occur, the density and quality of irrigation systems in rivers and fountains in the province where the base to develop the zoning map of flash flood risk is established, and the measures to prevent and mitigate the flash flood damages.

The execution work is to investigate and survey two main river systems, Srepôk, Krông Năng, and the current status of reservoirs, hydro-electric dams, and irrigation works in the province. Investigators are surveyors with professional skills, have made statistics on flash floods and inundation in two main river systems and recorded the current status of all irrigation works in Dak Lak province.

(c). **Content 3**: Assessing the current situation, analyzing the changes of factors that mainly affect the flash floods formation in Dak Lak province.

Approach, method, and implementation technique:

On the basis of basic data, plus identification of spatial and temporal variability (based on master plan), analyses identifies environmental issues, focusing on priority urgent, serious problems.

Study the human impacts on the component environments such as: Level of natural resource use, land use. Does the arrangement of industrial zones and urban areas ensure harm reduction due to flash floods? Check the quality progress of environmental components. From there, determining the main factors leading to the formation of flash floods.

(d). **Content 4**: Research and assessment of the current state of flash floods in Dak Lak province.

Approach, method, and implementation technique:

Based on the survey results of flash floods and landslides that happened in recent years, seven districts had many areas especially prone to those risks. Identify the main factors that cause flash floods in each region and pay special attention to the volatile factors such as the buffer surface: the vegetative cover and the surface geology are easily changed, especially in recent years, because of excessive exploitation of primeval forest, the planted forest area is not enough to compensate the exploited forest area; reclaiming land for production; developing settlements for free migrants without planning. Unsafe irrigation systems lead to flash floods when heavy rain causes dam failure. At the same time, analyze the information on the socio-economic development of each district in the coming years to consider whether it affects the formation of flash floods or not.

(e). **Content 5**: Analyzing the potential for flash floods based on closely related factors.

Approach, method, and implementation technique:

Flash flood formation is a combination of many factors with close correlation. Use the method of matrix establishing to evaluate the potential cause of flash floods by the following factors: Rain (precipitation, rainfall intensity, rainy time); topographic slope; cover level of the vegetation; river and fountain system (river grade, length of branches, river density, fluvial network density, runoff length); Soil environment (above is porous sand, below is tight clay or rock); Land use techniques, development activities related to land use and natural flow changes.

(f). **Content 6**: Simulation of flash floods evolution of some typical ones that happened in some main river basins.

(g). **Content 7**: Research to establish the zoning maps of flash flood risk in the province, (for example, Dak Lak) at scale 1/100,000 and zoning map of flash flood risk for key regions at scale 1/25,000.

(h). **Content 8**: Develop flash flood hazard maps for key regions and two main river basins at scale 1/25,000.

(i). **Content 9**: Develop an overall framework strategy for solutions to prevent and reduce damage caused by flash floods in the province (Dak Lak, Quang Ngai, and Binh Phuoc), period 2011–2020 with maps at scale 1/100,000 and scale 1/25,000.

(j). **Content 10**: Building flash flood potential database in Dak Lak province. Synthesize the databases and software for the topic:
  - Based on the maps of the content seven zoning map of flash flood risk at scale 1/100,000 and 1/25,000.
  - Based on GIS and RS software for research and warning of flood risks.

As the result, the study completes a flash flood risk map for Dak Lak province.

5.3.3   Scope of Research and Application of Hydraulic Model

5.3.3.1   River Network and Calculation

- **River network**

Existing topographic documents were used to determine the hydrological station network with records of water level, observed flow volume, river network limit, and hydraulic calculations of the river basin. For example, in Dak Lak, we take the Srepôk River, shown as follows:

  - Krông Buk river: From bridge 42 to Krông Pach river junction, length of 26,200 m.
  - Krông Pach river: From downstream of Krông Pach lake to Krông Ana river junction, length is 67,850 m.
  - Krông Ana river: From the confluence of Krông Pach river to the confluence of Srepôk river, 98,836 m long.
  - Krông Knô river: From Duc Xuyen to the confluence of Srepok river, 36,480 m long.
  - Srepôk River: From the confluence of Krông Ana and Krông Knô rivers to bridge 14, the length is 10,660 m.

The schematic diagram for the calculated river network is shown in Fig. 5.1.

- **Calculation**

Calculate the water level and flow volume at positions along the river network, determine the inundation level for the entire flood plain from Buôn Dray waterfall upward for the main flood calculation plans in October 2000, and plan to calculate early flood with frequency of 1% in Duc Xuyen.

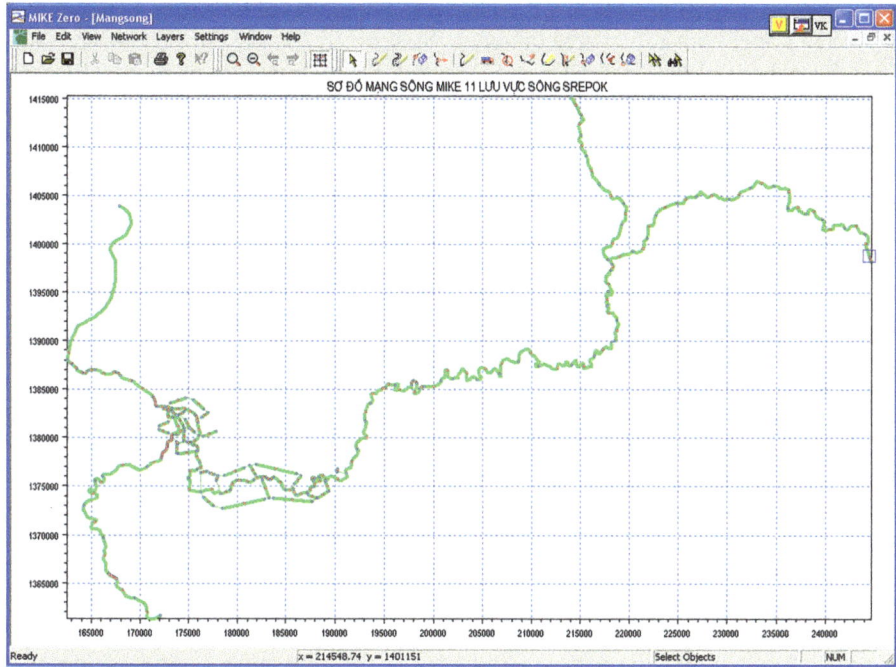

Fig. 5.1 Description of the river network structure of Mike 11 model for the Srepok River basin

- **The upper margin of the model**
- According to the river network, the upper margin of the model will be the process lines of the flow volume entering the positions:

  - Duc Xuyen on Krông Nô river.
  - Bridge 42 on Krông Buk river.
  - Upper Krông Pach line on Krông Pach river.

- **The downer margin of the model:**
- The downer margin of the model is the water level at bridge 14.
- **The margin along the axial river**

- In addition to the above margins, in the system there are about 4000 km^2 of catchment surface of fountains poured into the river network. Small streams will be gathered together for one entry point. Some main fountain branches joined are listed as follows:

  - On the Krông Pach branch from the upper Krông Pach line to downstream, there are fountains of Ea Dui, Ea Dram, Ea Pat, Ea Thu, Ea Dê, Ea Rok, and Ea Diê, with a total area of about 550 km^2.
  - On the Krông Buk branch, from bridge 42 to downstream, there are Ea Phê, Ea Kar, Ea Knăng, and Ea Uy fountains, with a total area of about 350 km^2.
  - Krông Bông branch has a basin surface of 808 km^2.

- From Krông Bông to Giang Son, including branches of Ea Krông Kma and Ea Sau Lang, with a total area of about 780 km².
- Branches of Lăk Dak Liêng and Dak Phơi have a total area of about 740 km².
- From Lăk to the confluence of the Srepok river there are Krông Diêt, Ea Pour, Krông Prum, and Ea Tul fountains, with a total catchment surface of about 140 km².
- From Duc Xuyen to the Srepôk confluence, there are branches of Dak Pri, Dak Drouk, and Ea Snô, with a total catchment surface of about 800 km².

5.3.3.2   Applied Mathematics Model

- **Model of application**

The MIKE model set of the Danish Institute of Hydraulics, including the hydrodynamic model MIKE11, was applied with the full high-level dynamic wave mode to calculate the one-way flow. Through application, MIKE11 has the ability to calculate with fast flow, rapid change of stagnant water efficiently and with a steep conductor-bed. Moreover, MIKE 11's interface is built on software of ARCVIEW, ARCGIS, being a GIS software has a nice interface and is very convenient for users.

After calculating, the results of MIKE 11 can be exported to MIKE 11-GIS to build the flooding maps to help users quickly analyze the flood conditions and the depth level of inundation, as a basis of flood damage analysis for basins. With these outstanding advantages, the MIKE 11 model has been selected for hydraulic calculation for the Srepok river network.

The flow regime for a single river distance is described by the system of differential derivative Saint-Venant equations (including continuity equations and momentum equations). The system of basic equations describes the hydraulic mechanism as follows:

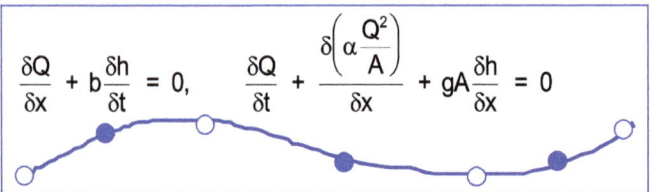

$$\frac{\delta Q}{\delta x} + b\frac{\delta h}{\delta t} = 0, \qquad \frac{\delta Q}{\delta t} + \frac{\delta\left(\alpha\frac{Q^2}{A}\right)}{\delta x} + gA\frac{\delta h}{\delta x} = 0$$

Solving the above system of differential equations according to the six-hidden-point finite difference method (Abbott-Ionescu six-point) will determine the flow rate, water level at every river distance, every cross section in the river network, and all timely points during the research period.

According to this method, the studied river network is divided into single river distances by horizontal cross sections. The river sections are connected in accordance with the natural state. The above method and the linearization difference process is used to obtain the system of differential equations written for the entire river network through the differential meshes.

Solving the system of differential equations will obtain the objective solutions at the meshes, specifically, to find the water level at the positions where there are cross-section and flow rate at positions between two cross-sections and at positions with works on the entire river network after each timely step of calculation.

Thus, after each calculation step, a flow value Q (m³/s) and the water level Z (m) will be obtained at the above-mentioned locations.

- **Algorithm for irrigation works**

Types of works: In MIKE 11, it is possible to include many types of works such as dams, bridges, culverts, pumping stations, and flood control reservoirs, and in each type of construction there are many work forms for simulation.

Internal conditions:
- The structures have the common internal conditions: Q = f (upstream/downstream water levels of the building).
- All works are located at point Q.
- Upstream/downstream cross-sections must be included in the database with the distance < dx-max from the project.

- **Calculation method for field plots (flood-plain)**

Based on existing contour and digital elevation maps, after going on field exploration to flood plain areas, consulting with people in the floodplain and studying the related documents, we find that the relationship between the floodplains and the riverbed is almost entirely natural. When the flood in the river rises, overflow will intrude into the floodplains, and when the flood comes down the water from the floodplains will withdraw to the river. In addition to the flood storage capacity, the floodplains also have the ability to transfer water.

Applied to the Srepôk river flood model because alongside of the Srepôk River there are the natural spilling beaches. When floods are high, they will freely overflow into the areas along two sides of the river. These beaches are not only capable of water storage but also capable of water transference. Therefore, we have separated the riverbed and the floodplains, and these floodplains are interconnected and connected to river cross-sections by branches.

Here we have modeled the Srepôk river downstream floodplains using 17 plots. These 17 plots are interconnected and connected to the Srepôk river cross-sections by 35 tributaries. The geometrical characteristics of these 35 branches were determined by testing the flood models in October 2000 and in August 2002.

5.4 GIS and RS Method for Zoning Map of Flash Flood Risk

The satellite image shows studied panorama. From there, the researchers from their knowledge can make a photo interpretation about the tone of image, color (often using wrong color combinations for composite images), shapes, and patterns or using software analysis to get the desired results.

In flash flood study, satellite images can give information about the current state of vegetation cover (vegetation index NDVI, topography (slope, side direction), rocky features, geological structure (crevices, gaps, folds), and even precipitation (meteorological satellites), which are very important factors in flash flood formation.

Interpretation of satellite images combined with field surveys will provide fast and accurate information about the studied area. These two needs are mutually complementary and supportive: interpreting images for an overview of the area and deciding points to be surveyed, and field surveys to help verify what has been interpreted and provide what has not been known when interpreting the image.

The study of flash floods using satellite images of ground observation will help to determine the features of the area, and depending on the spatial resolution of the satellite image, it can be determined at different levels of detail. From there, it is possible to combine with GIS to map the vegetation distribution, soil and rock distribution, topographic elements, etc. These are the basic information sets to establish a flash flood zoning map to help projectors of flash floods to easily forecast the annual flood season.

Briefly, the most important function of remote sensing in flash flood research is to provide relatively accurate data on the current state of land use, factors of terrain, and geology from satellite and aerial imagery through analysis and interpretation of images by software or researchers. These are necessary data to build a zoning map of flash flood risk. Moreover, these satellite images can be regularly provided by image reception and photography stations. Thus, it is possible to successively update the current state of the area, especially for factors that easily change such as vegetation cover.

All research results in this chapter are intended to build the map of flash flood risk for Dak Lak province.

5.5 Conclusion on Zoning Map of Flash Flood Risk

Many areas in Dak Lak province have natural conditions with potential risk of flash floods in the rainy season such as heavy and concentrated rainfall, big steep gradient, narrow fountain/ riverbeds, shallow streams, and thus water can gather very quickly. At the same time, the soil and rocks on the slope and along the valley are porous and weathered easily to form pebbles, so the flood flow has a lot of solid matter forming a rocky-mud flood. Vegetation cover along valleys, alluvial flats, low-lying terraces affected by resident people tend to decrease, limiting the effect of preventing floods.

The application of geographic information systems GIS allows building digital elevation models DEM to determine the parameters of slope and valley shape that indicate flash flood risk. The satellite image interpretation allows calculating its cover level and determining the density of distributed plants. Combining this information with the maps of soil and rock types allows the establishment of zoning map of flash flood risk in Dak Lak province.

In the research scope of this topic, areas of permanent flash floods occurrence are found in Dak Lak province of Central Highlands. Stemming from the topographical characteristics and river network distribution as well as related characteristics in the province of Dak Lak, the flow feature is quite different. That is, some areas only have normal floods and inundation, while some places have frequent flash floods.

Considering the results and survey together with the synthetic analysis, we assess that Dak Lak province has five areas prone to flash floods, in other words, there are five main areas at very high risk of flash floods.

Zone 1 Mostly includes Cư Mgar district (area with very high risk of flash flood, accounting for 14.72%) The catchment includes fountains Ea Tul and Ea Huar. Starting from 700 m upstream (Cư Dlie Mnông commune) flowing down 200 m downstream (Ea Huar commune), the flash flood area includes communes: Cư Dlie Mnông, Ea Tar, Ea Kpam, Ea Hđinh, Ea Kiet, Cư M'Gar, Ea Mroh, Quang Hiep, and Ea M'năng (belonging to Cư M'Gar district). Ea Huar, Ea Wer (Buôn Dôn District).

Zone 2 Mostly includes Krông Năng district (area with very high risk of flash flood, accounting for 12.45%) of Krông Năng river basin. Flowing from 750–800 m upstream, to 400–500 m downstream, through the communes of Dliê Ya, Cư Klông, Ea Tam, Phu Loc, Tam Giang, Ea Puk, Krông Năng Town, Phu Xuan, Ea Dah belongs to Krông Năng and Xuan Phu districts, Ea Kar town is in Ea Kar district.

Zone 3 Includes M'Drăk district (with 3.98% of the area at very high risk of flash flood) Krông Jing Fountain, originating from an altitude of 500 m in Cư M'tar and Cư Kroa commune, falling to an altitude of 400 m in Krông Jing commune and M'Drăk town.

Zone 4 Includes Lak district (4.98% of the area at very high risk of flash flood) stretching along the fountains of Dăk Phơi, Dak Liêng of Lăk district. Upstream with an altitude of 1000–1200 m (Dăk Phơi commune, Bông Krăng) flowing down to 400–500 m altitude in two communes of Dăk Nuê, Dak Liêng and a part of Dak Liêng town.

Zone 5 Belonging to Ea Hleo district (8.80% of the area with very high risk of flash flood) Ea Hleo district communes: Ea Tir, Ea Khal, Cu Amung, Ea Wy, Cư Môk, and Ea Drăng town.

Based on the actual conditions of Dak Lak province – highly divided topography, limited vegetation, high annual rainfall, and dense irrigation system – the research team has proposed appropriate solutions:

- Construct the upstream water-regulating works and build firm reservoirs for both irrigation and flood control.
- Upgrade and repair the degraded irrigation works.
- Increase the flood drainage capacity of the conductor-beds.
- Land schematic planning in the basin for avoidance of flash floods.
- Use land reasonably.
- Afforestation and forest protection.

5.6   Recommendations on Dak Lak Flash Flood Measures

The products have been elaborated completely and in detail. For maximum effectiveness, products need to be disseminated to the relevant authorities. For places prone to flash floods and near residential areas, there should be a migration plan or urgent construction of response facilities. At the same time, local authorities also need to have a plan to regularly update the database on flash floods in Dak Lak.

Flash floods and landslides are natural disasters that frequently occur and are difficult to accurately forecast. Therefore, it is suggested that authorities of all levels consider the evacuation to prevent natural disasters as an important task, regularly paying attention to it and directing the integration of measures to prevent flash floods into the local activities. It is necessary to review and assess new critical areas. The migration program should be integrated with the national target program on new rural construction, with priority given to urgent areas.

For the recommended comparison, we can refer to the project of a hydro-meteorological multi modeling system, made by Asst Professor Binata Roy and Prof Dr. AKM Saiful Islam at the Institute of Water and Flood Management (IWFM), Bangladesh, proposed for five northwestern districts of Bangladesh covering highly flood-prone areas. The three-staged multi-modeling framework used in this project was developed using a high-resolution and open-source modeling system, which makes it economic, easily accessible, and operational for both high-tech and medium-tech, for official or personal application.

At first, a numerical weather prediction system is used incorporating the terrestrial and gridded information and the cloud physics available from various world weather forecasting centers (GPS, ECMWF, etc.). Then the rainfall and temperature output of the weather model is used as the boundary input of a hydrologic model as a rainfall–runoff model. Later, the hydrologic model output moves to input of the hydrodynamic model as a streamflow routing. The final output of the hydrodynamic model – the water level – is later compared with the existing danger level determined by the Center of Analysis and Forecast (Fig. 5.2).

Using this modeling project, the flood forecasting can be conducted by different local forecaster or other NGO specialized centers. Institute of Water and Flood Management (IWFM), Bangladesh, is expecting to develop this system for proper forecasting and early warning to communities and also plan another full-fledged monsoon flood forecasting and warning tool before the start of the next monsoon in Bangladesh. This a very optimal and suitable model of flash flood forecast for condition of Vietnamese mountainous regions. Vietnam can additionally use this modeling system for the flash flood forecast in Dak Lak or other provinces with similar risk.

Currently, there are many reservoirs remaining after exploitation periods that have seriously deteriorated and not been repaired, threatening to become unsafe in the rainy season. Reservoir dam failure is a major cause of flash floods. Therefore,

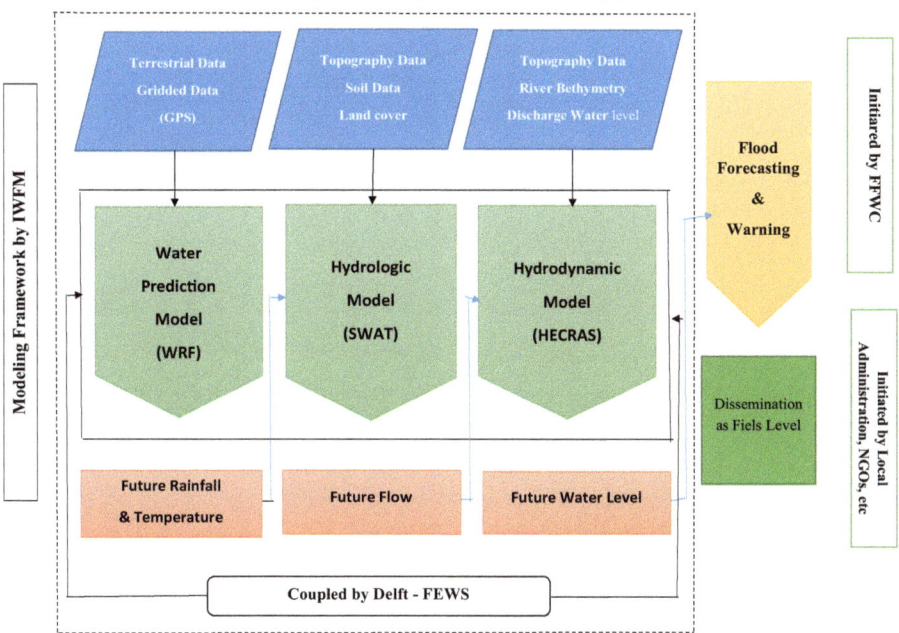

Fig. 5.2 The 3-staged multi modeling framework is being used in this project

it is suggested that the provincial People's Committee should have a plan to fund upgrading the unsafe irrigation works of water storage. In addition, the inter-lake systems suggest that the construction owners need to coordinate to build the process of operating the inter-reservoir reservoir to ensure safety when regulating flood discharge.

The forest area in the province is being significantly reduced, especially the watershed protection forest, which is also a factor contributing to the increase of flash floods and landslides in terms of both frequency and intensity of destruction. In order to limit the damage caused by flash floods and landslides, the Provincial People's Committee should pay attention to the implementation of the plan for afforestation and protection and pay special attention to the protection of the upstream forests of the province. Annually, the PPC is requested to direct the investigation and review of the remaining areas of the province to develop a reasonable and timely prevention plan, especially to build a plan to migrate out of dangerous disaster areas.

Chapter 6
Solutions Against Flash Flood Hazards (Case Study: Binh Phuoc Province)

Abstract This chapter presents the measures to prevent and limit the harmful effects of flash floods in mountainous areas of Vietnam in general and Binh Phuoc province in particular (as a case study). The chapter outlines the specific ways of reaction for each person when facing flash flood – from before the flood comes, handling situations during the flood, and dealing behaviors after the flood, and outlines the immediate handling measures with appropriately basic methods for the Government in responding to flash floods.

The chapter proposes structural and non-structural solutions to prevent flash floods. The structural works include to increase the flood drainage capacity of the conductor-bed, divide the flood currents, separate the mixed solids from the flood mass, use irrigational works, reduce the damage possibility to the works, have a plan to regulate the water of lakes and dams, and arrange for land use guided by science. The non-structural works include good planning of land and managing organization, arranging the appropriate plant allocation and crops, afforestation and protecting forests, building an early warning monitoring system, and teaching the behaviors of environmental protection and disaster response.

Finally, some concluding paragraphs discuss all that has been presented in the monograph.

Keywords On-site avoidance · SALT model · Installation and trial · Artificial riverbeds

After many years of research on the mountainous regions of Central Vietnam, the South East, and the Central Highlands (Provinces of Binh Phuoc, Quang Ngai, Dak Lak), we have synthesized and summarized the solutions to prevent and limit the harmful effects of flash and tubing floods in these places, including structural, non-structural, and management solutions. This chapter chooses Binh Phuoc province as a representative mountainous region as a typical example for the application of the above solutions.

© The Author(s), under exclusive license to Springer Nature Switzerland AG 2022   119
L. H. Ba et al., *Flash Floods in Vietnam*,
https://doi.org/10.1007/978-3-031-10532-6_6

6.1   Summary of Specific Solutions

6.1.1   Solutions for On-site Avoidance

Many narrow strips of coastal lowlands and the mountainous-hilly lands are not strangers to flooding because these areas always face challenges from rising river water and tropical storms from the sea, as well as too much rain from concentrated rains. Humans that live or work in flash flood prone areas need to be equipped with basic knowledge and what to do before, during, and after a flood.

(1). **Before flash flood – Always be ready**

Being ready is the best solution for offices, units, and local people to stay safe and protect property when there is a flood. The local government at all grades should flexibly deploy the plan of preparing on-site forces to be prepared to deal with flood hazards with sufficient material–human resources for the community. Each person takes these simple steps to be prepared for whatever the weather causes.

– *Distinguish the flash flood warning from flood forecasting*

Flood monitoring simply means elaborate observation on favorable flood conditions and informing people of predetermined forecasts. This is a good time to check the emergency kit and prepare any necessary items to bring in the event of an evacuation, such as clothes, medicine, and food. Follow the weather reports and guidance from specialists and authorities. Whereas Flash Flood Warning means flash flooding is happening or imminent, and **when warning comes, everybody has to evacuate immediately** because the event happens very quickly and there is little time to react.

– *Preparation for houses, offices, works*

Reinforce houses, constructions, cover sand bags, anchor roofs, brace pillars, and create an emergency escape exit, if there is time. This can help prevent floodwaters, storms, and winds from sweeping away everything. Ensure that the electricity and water system, stove, machines, and equipment are well protected from serious damage and danger.

– *Prepare an available emergency kit*

Keep important documents bag, food, field rations, bottled water, life-jacket, plastic sheets, medicine, and everything needed for 3 days in case you get stuck in place. The kit also includes a battery-operated radio, removable battery, a portable phone charger, a battery torch, a small knife, an opener, a swimming float, a lighter, and first aid box. If possible, add a pocket compass, rope coil, sleeping bag, etc., as possible.

– *Urgent evacuation*

No awaiting for an evacuation order. If a flood is imminent, let's rely on official weather reports and use your best judgment early to move to a higher place and take immediate action to move. If waiting for the evacuation order or lingering to calculate and discuss, it is possible to be stuck at the place and cannot get out of the flooded area on time.

(2). **During flash flood – Always take precautions**

A flood situation is an emergency that requires attention and care to detail in order to avoid injury. It would not be in vain if everyone knows how to behave in this coming disaster

– *Come to higher place*

Absolutely do not hesitate but must act decisively at-once. If possible, immediately leave the lower land for a higher place. If the road is flooded, find your way to the roof of the house and call for help from your cell phone or any other means possible. Absolutely do not dwell on the bare hillsides, the geological base does not seem to be firm and generates strange noises from the ground (Fig. 6.1).

– *Avoid flood water*

Stay away even from an area that is not too deep or dangerous. Just 20 cm of rising water can make it difficult for people to move. From 40 to 50 cm of water can lift and carry even the biggest vehicle – like a truck. Drowning is not

Fig. 6.1 Escape flood in the Vietnam Central region

the only danger to worry about; floodwaters are often polluted by raw sewage, garbage, and chemical runoff, causing all kinds of dangerous diseases.

– *Avoid electricity*

Low electric wires and submersible or wall outlets can electrify the flooding water in houses and transmit within a certain radius. If you are not sure that the power supply has been disconnected, do not enter a potentially electrified area.

(3). **After flash flood – Wait patiently**

After the floodwater recedes or the storm dissipates. Everybody wants to get life back to normal, but in some cases, it will take a while for everything to return entirely under control.

– *Wait for all to be clear*

Do not enter any location damaged by flash floods, inundation, or landslides until it is declared safe by the rescue unit or local authorities. Water can weaken the structural supports and contribute to the development of hidden mold. Flash floods can empty the guts of the building foot and loosen the porous ground and the area can sink and bury, suddenly collapse, difficult to sustain or escape on time.

– *Avoid hurt areas*

If one must return to the site of damage or flooding in the area around to survey the damage or assist the remedial efforts, stay away and wait until the dedicated rescue force opens the way to there (Fig. 6.2).

Fig. 6.2 Real scene after flash flood

– **Contact the recovery service**

As soon as we can return to our home or facility, is the right time to contact the insurance and a professional service that recovers the damaged properties from water or storm. If we repair and restore infrastructure ourselves, we should also create a safe site, full of tools and preferably with the support of neighbors or the community.

– **Follow information after storms – floods**

Even if everything can be restored enough temporarily, although not as it is previously, it is still necessary to regularly follow the media news, weather forecasts and the instructions of the specialized agency or local authorities to avoid storms and floods or ensure safe conditions for living.

6.1.2   Governmental Solutions

Many flood control projects have been implemented by the government for many years in mountainous and midland provinces with potential flood risks. The government's specialized hydro-meteorological and disaster prevention have also organized systems to monitor, measure, synthesize, analyze, warn, and operate in emergency situations. It is now necessary to fully complement the solid and flexible flood infrastructure networks in critical areas and strengthen the capacity to monitor, analyze, forecast, and warn of flash floods – reasonably appropriate under the limited conditions of our resources (Fig. 6.3).

Fig. 6.3 Stone in galvanized net armor for embankment against floods in Vietnam

However, overall, the extremely aggressive quickness of flash floods, and the tragic reality of the recent great floods, make us see how much we have invested in avoiding and fighting, and the current ways are incomplete to withstand the force of nature. The problem must focus on flood prevention with structural works for flood control – diversion, flow regulation, shelter and settlement areas, and above all, an early warning system. In this spirit, the Government is urgently reviewing the proposals to allow the implementation of the Project of Installation and Trial of Early Monitoring/Warning System and protective dams against rock–mud flash floods to mitigate natural disaster risks caused by flash floods and landslides at some high risk sites (in the period 2020–2025).

In order to mitigate the damage caused by flash floods, some measures must be taken:

- Having a drastic will and policy suitable to the context of climate change in the short and long term with a vision to +20, 30 years.
- Completion of flood control works in key points, densely populated areas with reasonable investment, adequate and regular inspection, maintenance, and upgrading.
- Education to raise the awareness of environmental protection, the method to identify natural hazards and adverse impacts on the environment, the method of handling situations, effective response actions in natural disasters, and a spirit of mutual support for the community.
- Enhance the improvement of the capacity for climate forecasting, especially the early warning of flash floods with well-trained and experienced professional personnel, full of advanced equipment with high, modern technology content. Accelerate updates and upgrades to the forecasting technology based on new scientific and meteorological achievements such as flood generation modeling technology, flood dynamics analysis algorithm, wide spectrum and high resolution satellite image interpretation, and remote sensing techniques.
- Build an interactive data network of industry, national, regional, and international to transmit, synthesize, process, and compare toward a homogeneous and mutual action. Build an effective communication system to communicate events, educate the public awareness and action, and enforce forecasts, warnings, and implementation guidance.
- International cooperation in any problem related to floods and natural disasters, such as the exchange of information, data, communication, training, system upgrading, sharing of experts, technology, equipment, companionship providing relief, sharing experiences, adding forces aimed to modernize, internationalize extensively increasing the specialized management activities, and handle emergencies.
- Semi-professionalize the locality official service force to undertake propaganda, information, warning, rescue, and remediation. This force is the core of the "reaction on-site " motto and always at the forefront in emergency situations.
- Support with higher, more appropriate resources from upper authorities for local efforts of forecast communication, intervention action, response aid, rescue,

distribution of necessities. This support often requires the activities of the Military forces, the Police, and the higher government departments.
– Available process for promptly responding to cases of sudden flooding or great damage caused by natural disasters. Having the adequate and effective policies and sanctions versus the acts of harming the environment of individuals, groups, units, and enterprises. To protect the country for sustainable development in the spirit of "not exchanging environment for economic efficiency and GDP growth."

6.1.3   Radical Solutions for Preventing Floods

With 30% of people living in mountainous, coastal areas vulnerable to flooding, Vietnam needs to react creatively so that other countries can see the lessons learned; otherwise it will be hit by successive flood disasters in the future. When continuous heavy rain on highland or sea levels rise on coastal land, floodwaters will rise higher, quickly concentrate, becoming a catastrophic flash flood. These areas will deteriorate further each year, so Vietnam needs to consider some reaction options.

The chosen options include building the defense walls, finding the resettlement sites for people or a new livelihood for them, and replanning the rural and urban areas. At the same time, Vietnam should promote the long-term flood warning system based on the progress of satellite image observation, geographic information system GIS, and digital algorithms to help people evacuate safely.

In the urgent choice to pay attention and invest between the monitoring, warning, and residential planning system in flood-sensitive areas, it is suggested that the chosen priority be the residential and production planning solutions because it is easy to concentrate resources to stabilize life and avoid consuming too much financial resources for equipment because the early warning monitoring facility and system is an expensive investment – of course, it will be implemented but must be done later by itinerary (Fig. 6.4).

Recent floods and landslides have occurred in Thua Thien – Hue and Quang Nam, while other districts in the two above provinces with loss of people and property, but in Tay Giang district of Quang Nam alone. Due to the policy and planning from a few years ago, the settlement of about 90 ethnic villages was relocated to safe isolated areas from steep slopes, along rivers and streams, so they were able to escape the disaster but still in stable production for a long time.

Accordingly, the radical flood control facility needs to form in two solution groups – structural and non-structural:

– **Structural solutions**

   • It is essential to protect primary forests and develop secondary natural forests to consolidate the watershed protection lines with forests and natural vegetation. Therefore, it is necessary to plant perennial trees, solid timber trees, multi-layered & multi-species trees, and densely spread vegetation with close-to-soil layers, vines, and grass with thick deep roots in order to develop the natural

Fig. 6.4 Sluice to regulate flood discharge in Nghe An

biodiversity. In addition, develop should not expand for the planted forests of industry, fruit, perennial trees, or green cover plant vegetation in upstream areas because they do not have any suitable effect but cause further harms to the forest shelf and add "material toys" to the attack force of flash flood flows, and when the flood recedes, make the scene more disordered, disturbed, and dirty.

- Completing the network of dikes to prevent and divert floods is very effective in key areas to maximally limit the concentrated, tubing, resonant floods and diversely spread the flood flow in more directions and exit ways, to reduce the destructive power. To improve the slope of critical points, build walls and dykes to prevent slip and landslides. Install drains or flood control sluices at the gates open to the downstream basin (Fig. 6.5).

- Construct the public works and special infrastructure that can be coordinated to prevent flash floods. Design and construction of mobile houses in the form of "anti-flood, follow flood"… in order to survive in flash flood conditions when the escape route or traffic is congested or destroyed.

- Semi-firm and firm works in the area must take into account geological bases, improve slope, reinforce the works sustainably with pre-forcing effects of flash floods and the generation and development of flash floods. Must fully assess environmental impacts and resilience to natural disasters.

- Plan the residential areas to live safely with floods in order to choose a geographic area that is sustainable, low-risk, and the separated production and service land near that place with lower safety. Construction of flood-plain settlements requires compliance with rules that are flexible to extreme weather and flooding, wildfires, logistical conditions, on-site medical care, and security. Permanent residence (Fig. 6.6).

Fig. 6.5 Construction of dykes to prevent floods in Song Dinh

Fig. 6.6 Vietnam anti-flood house

– **Non-structural solutions**

- Build the zoning map of flash flood forecast and warning with appropriate scales and treatment hierarchy; profit by scientific and technical advances to perfect the early monitoring and warning technology. Due to the extremely fast nature of flash floods, it is necessary to have early warning, using the collection and interpretation technology for satellite images, the geographic information systems (GIS), remote sensing (RS), development of modeling and analysis algorithm, combined with the available hydro-meteorology forecast monitoring system to give the accurate pre-flood evolutions and timely information of forecasts and early warning.
- Build an appropriate communication network to promptly rescue and act as an avoidance guide. Emergency warning information by loudspeaker, telephone, text message, or flare signal must be sent to the receiver at least 1 h before a flash flood, the early warning message must come at least 3 h before the event.
- Create an online transmission line to observe, monitor, and process the reservoir information and water regulating activities through sluice gates, valves, and spillways of all hydro-power dams, the natural lakes with regulating gates, the ecology, irrigation or water supply lakes from large to small and medium. The program and practice of water regulation of these objects must be 100% through the inspection and coordination of the specialized management unit. All reservoir alarms and flood discharge intentions must be notified and approved by authorities.
- Build immediately on this disaster-sensitive area an appropriate weather measurement network to monitor and forecast the rain in the flash floods basin with operational staff with fixed or mobile equipment. More advanced, more professional, higher technology content to deliver the precise and timely forecasts.
- Propagandize and promote the community to understand and prevent this type of natural disaster. Organize some workshops in the community to train coping skills, take urgent action in the flash flood situation, and deploy a force of volunteers scattered throughout the areas.
- Organize a flood prevention, treatment and rescue system under the motto "4 on-site" (forecast, coordination, command/response force/material facilities/ logistics, overcoming) to react quickly and effectively in areas affected by natural disasters (Fig. 6.7).

6.2   On Structural Solutions

The structural solutions are carried-out to reduce the scale and impact of flash and tubing floods. They are often an optimal combination of solutions and also incorporate the non-structural solutions to mitigate flash floods themselves and their harm

Fig. 6.7   Armed forces help people during the flood

levels. The application of a rational combination in flood control is considered from the perspective of reducing the loss of people's lives and property. Each combination corresponds to certain natural and social conditions. This section will take what related to flash flood control and prevention in Binh Phuoc province, Vietnam, as a typical example.

6.2.1   Increase the Flood Drainage Capacity of the Conductor-Bed

Bad drainage is an important factor influencing flood intensity. Especially in the downstream sections, when floods gather quickly, with large flow but bad drainage, so their impacts will be stronger. The causes affecting flood drainage mainly include narrow, winding, or households encroach on and throw garbage, creating obstacles blocking the flow. In Binh Phuoc province, localities situated in high risk of flash floods and with narrow flow conditions, winding, or encroached by people include:

Phuoc Long district: Tributaries of Dak Rat, Suoi Rai, Dak Ran, Dak Lai. Dak O, Dak Ken Roi, Dak Kang, Be river, Thac Mo lake, fountains of Dam, Driem, and Dak, Dak Prodan.

Bu Dang district: Da Trio stream, tributaries of Dako, Da R'Lon, Dak R'Lon, DaR'Dê, Dak K'Meu, and Dak Glun river.

Dong Phu district: Fountains of Plee, Nhung, Rua, Bang, Rat, Bon, Cam, and tributary of Pa Pech (Bu fountain).
Bu Dop district: Tributary of Be river, fountains of Chum Dien and Phao.
Binh Long district: Fountains of Xa Cam, Cha La, Xa Cat, Don, Heo, Suoi Ru, Cha La Stream, Ro, Lai, Cat, and tributaries of Chum Ri, Ao No, No Nong, and Bu Dinh river.
Loc Ninh District: Fountains of Das, K'Lieu, Cham Ri, Ngom, KhLey, Trau, Can Reng, Thao, Lovêa, Yor, M'Kei, Bresson, and tributaries of Brô Sinh, Tam Buo, Bong Cam, Chang Roai, and Ton Le Cham River.
Dong Xoai town: Fountains of Che, Rai, and Num.

In order to prevent and mitigate the effects of flash, tubing floods, the above localities need to regularly carry out the following activities:

- Destroying, removing natural obstacles: clearing trees in the conductor bed area.
- Eliminating artificial obstacles such as bridges, damaged works, solid materials piled up in the conductor bed, blocking the flow, unreasonable construction works.
- Regulating methods of exploiting materials on rivers, in the conductor bed, residential areas, removing areas encroaching on conductor beds, river banks, and blocking flow. These works require regular execution before each rainy season and during the flood season.

6.2.2   Splitting Flood Flow

Splitting is action to reduce the severity of the flood (time, intensity, flood peak) thereby reducing the impact of destruction and sedimentation. Flood splitting is to cause a part or all of the flash flood flow (both liquid and solid) to follow a different route to the main river or catchment zone so as not to damage the protected area in the valley or basin of the river.

Areas at high risk of flash floods, characterized by sparse, meandering waterways and near large rivers, are very suitable for this solution. Meanwhile, the splitting is done by digging a canal to lead to the main river or low-lying area.

6.2.3   Separate the Solid Matter from Flood Flow

This solution is to separate water from solid matter (crushed rocks, boulders, pebbles, gravel, big trees, etc.) in the flood flow to reduce its impact force. Once most of the solid matter has been separated, flash floods will become normal floods and can be mitigated by habitually fluvial structures. For separating solid matters, dams with a small bottom slope or low barrier dams at the conductor-bed are often used (Fig. 6.8).

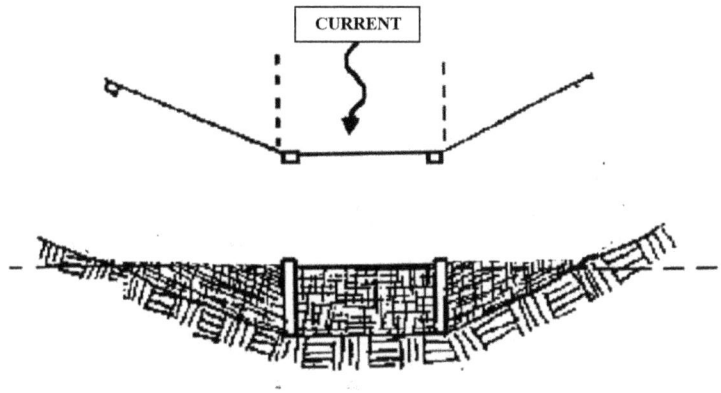

Fig. 6.8 Construction of low barrier dam

6.2.4   Irrigation Technical Solutions

Irrigation technical solutions to improve the mountainous terrain, interrupt flow, store water in steep slopes, and implement irrigation. This is an important and very effective measure for flow control. When implementing the structural solutions, it is necessary to base them on reasonable land use planning in close combination with other solutions. The simple structural solutions include embankment to retain water, dig ditches to prevent water in streams, and build the control dams. Other important irrigation solutions are the construction of flood control reservoirs, flood retardation works, and flood drainage.

- **Covering water in the slope**

  This is a basic and single solution to prevent erosion and flash flooding on sloping land. On a gentle slope, light soil, with good or medium water permeability but not very large currents, it is possible to build steep horizontal banks along the contour lines, on the trunk and the shore surface for grass growing. The shore is 34–42 cm high, and the bottom of the bank is 1.0–1.5 m wide. It can be done in the form of soft banks (not tightly compacted), or half-hard banks (deck ground is compact but not the upper layer). If heavy rains, the grass banks that do not hold all the water can still overflow the shore (ditch on the lower bank) or dig a reservoir at the edge of the bank to increase water retention (Fig. 6.9).

  Suitable areas for this solution are:

Phuoc Long district: Thac Mo lake; Tributaries of Dak Tang & Be river; Dak Ó, Hamlet 7 of Dak Kang branch; and Dak Ó, Dak Glun, Dak Me.

Bu Dang district: Da Trio fountain; Tributary of Dako; Hamlet 2, Dako, Da Bo; Da R'Lon branch; Da Nao, Da War; Tributaries of Dak R'Lap, DaR'De, Dak K 'Meu, Dak Glun, and Dak Mơ.

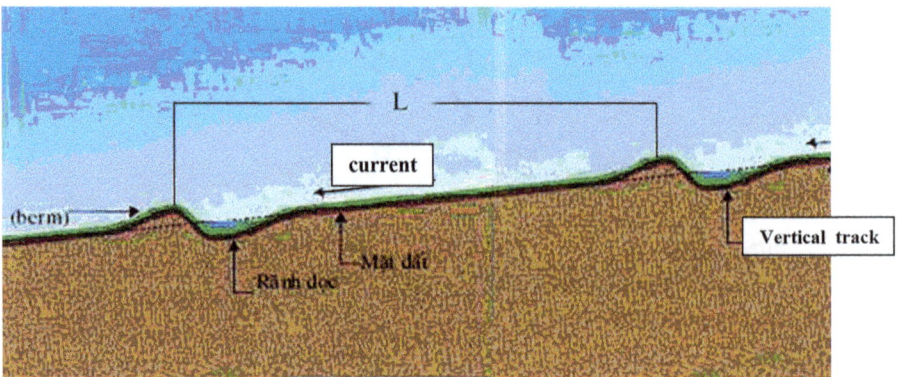

Fig. 6.9 The barrier holds the water on the slope

Dong Phu district: Fountains of Plee, Nhung, Mung, Tre, Rua, Bang, Nghia Hoa, Rai, Bon, Cam, Tributary of Pa Pep (Bu fountain), and Hamlet 4.

Bu Dop District: Tributary of Be River.

Binh Long district: Fountains of Don, Heo, Hung Lap A, Ro, Cat, tributaries of Ao No, Chum Ri, Bu Dinh river, Lac swamp, villages of Trao, Bung, and Nui Gio 1, 2, Hamlet 7.

Loc Ninh district: Fountains of Trau, Cham Ri, Ngom, Thao, Lovêa Yor stream, M'Kei, tributaries of Bro Sinh, Chang Roai, Bong Cam river, and Hamlet 6.

- **Install dams to prevent water in streams**

In the case these solutions do not obstruct the water flow, it is necessary to use measures to install dams in the streams. Most of the rainfall and surface currents on the slope are concentrated in the streams, so if the flow is prevented, the flood flow is limited, owing to effects of the dam, as below:

- Reducing the flow rate in the stream and its erosion, thus reducing flooding for the downstream production area.
- Prevent mud and sand, reduce flow rate to bring alluvium to rivers, and create conditions to profit by streams.
- Limiting the expansion and deepening of the stream.

Dam can be made with earth, stone, or wood; this measure is effective when combined with other flood control solutions, especially the construction of protective forests and small and medium reservoirs in downstream fountains and small rivers.

6.2.5 Mitigate Flood Damage for Traffic Works

Flash floods often cause great damage to traffic works such as broken bridges, collapses, and missed roads, causing traffic congestion; therefore, it is necessary to take the solutions to protect these works.

Fig. 6.10 Deepening
artificial riverbed

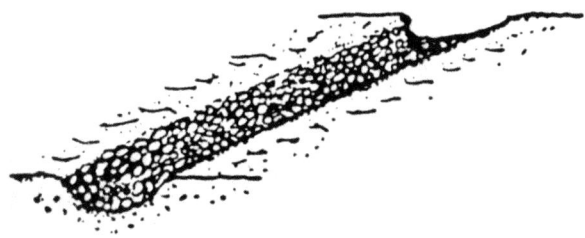

- **Widening the aperture of bridges**

Widening the aperture of bridges and culverts is an especially necessary measure in river sections in frequent occurrence of flash floods. The accretion created by flash floods is increasing, the bridge aperture will not be enough to drain water, and it will need to be widened or more bridges built in the low position of the accretion. To do this, it is necessary to determine the characteristics of flash floods, especially their flow volumes across the bridge.

- **Arrangement of bridges and works**

Principle of Bridge/Road Arrangement
- For roads passing through flash flood areas, to ensure the safety of road bottoms and bridges, additional funding for maintenance and construction of protective works should be increased. At the same time, it is necessary to build a plan the avoidable lines outside the flash flood flow.
- When choosing the bridge location, it is necessary to consider the investigation documents of the specific situation and parallelly arrange diversion works properly, and have appropriate protective measures for bridges and road bottom.

When the slope of the riverbed is small, there is little sand and impurities in the water stream, but the flow is large. If we want the water not to overflow the accretion section, endangering the road bottom, we can make an artificial riverbed to adjust water flows into the bridge, and pay attention to extend to a certain section downstream to be careful of the phenomenon of sudden expansion flow on the downstream bridge. Artificial riverbeds must be straight or gradually curved to facilitate the flow of mud and rock to gradually flow into the bridge (Fig. 6.10).

Principle of Determining the Bridge Aperture
Because flash flood flows carry a large amount of boulders, mud and sand, the dynamic balance velocity is greater than normal water flow; therefore, it is easy to cause much harm to the road bottom and bridges. When determining the size of the bridge aperture, the static space under the bridge should be sufficient to ensure that objects caught in the flow of water easily move through the bridge. Usually the following points should be noted:

- On places of drifted rock mud, the bridge layout is better than the sluice; bridge aperture should not be narrow, must be equal to the width of the natural riverbed corresponding to the design water level at the entrance, and at the same time should choose a one-span bridge.

- Static height under the bridge must be enough. When surveying the fluctuation rule in sedimentation and erosion of the river, it is necessary to consider the average sediment height for many years and the height that can be maximally deposited for each time; choose the maximum value to calculate the designed water level.

6.2.6 Limit Flash Floods with Control Dams

For areas where the forest area was strongly destroyed such as Binh Phuoc province, the catchment surface was exploited, destroyed, weathered, meaning that there are many favorable buffering conditions of flash floods, the control dams allow effectively controlling the sediment and runoff for reducing wild flash flood in particular and every flood in general. Control dams are usually very small, similar to water catching ledges in mountainous areas. These low dams are built on rivers and streams to store flows.

Control dams are usually a simple structure, made of gravel and clay to form a closed dam. The designs of these dams vary widely depending on the typical topography and construction site. The dam is easy to build, fast and cheap, making it suitable for rural mountainous conditions. The series of these dams on the main river or in tributaries allows maintaining flash floods flow, keeping mud, rocks, and solid substances in the flood. So the dams help to reduce basin erosion, flow velocity, and obvious amount of mud and rock in the downstream flood of the dam. The dams also have the effect of regulating the flow of the season, year, and maintaining the soil moisture in the dry season (Fig. 6.11).

6.2.7 Reasonable Land Use

In Binh Phuoc, up to 18.54% of the population are ethnic minorities with ineffectively traditional cultivation methods that cause many negative impacts on mountainous land, thereby increasing the risk of flash flood formation. Therefore, it is very important to know and implement good techniques in land use.

6.2.7.1 Land Preparation and Reclamation

Soil preparation before sowing is very important in reducing erosion and runoff. Plowing must be carried out precisely in a contour line, evenly and deeply.

- **Deep plowing as per contour lines:** This is an important measure to create many small trenches across the slope; each furrow acts as a water barrier, making rainwater more retained. Moreover, the depth of the plowed soil will increase porosity, so the water permeability and retention capacity of the soil is also enhanced,

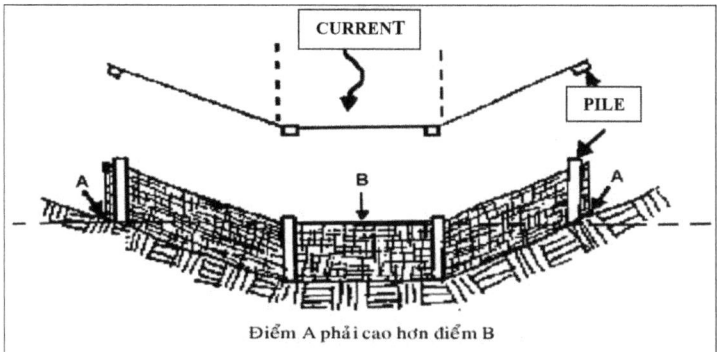

Fig. 6.11 Dams control

thus limiting flow. However, in places with high slope and heavy rainfall, where tight soil has difficultly absorbing water, the effectiveness of this measure is not great, and needs to be combined with other measures.

- *Soil-bed earthing as per contour lines:* On the basis of deep plowing on contour lines, soil-bed earthing on contour lines is effective against erosion and great currents. Contour soil bed can reduce the flow rate by 60–90%, reduce the amount of soil erosion by 80–95%, and increase the yield by 8–33% compared to non-soil bed land. The advantages of making horizontal beds are to improve the terrain, increase the rain-catching area of the ground, decrease rainfall per unit-area; Each soil bed acts as a water barrier to cross the flow, the water that cannot absorb enough will be pushed down to keep in the middle of two beds and then continue to seep into the soil. Moreover, the soil beds increase ground porosity and permeability.

6.2.7.2   Cultivation Techniques on Sloping Land

- *Planting in rows on contour lines:* is a measure that aims to prevent and mitigate the flow rate, increase the water amount seeped into the soil, thereby reducing the soil amount washed away, increasing crop yields. Actually, this measure is relatively popular in Vietnam and is one of the principal farming techniques on sloping land.
- *Inter-planting with overlap crop:* is the long-standing experience of Vietnamese people and also widely applied in other tropical countries. Intercropping is a measure to make the most of the productivity of the site conditions, and parallelly, has the effect of covering and improving the soil very well. Over lapping crop is a method that makes the ground always covered with plants, harvests many products in a short time, protects arable land from erosion, and limits flow. In the intercropping method, attention should be paid to intercropping the agricultural with forestry plants (agriculture/forestry coordinated). This is an effective measure, widely applied by the people.

- ***Inter-planting tree rows on contour lines:*** is a measure to effectively prevent flow and erosion then increase crop yields. This method divides the slope into many sections, with each one thickening up the grown trees to another one with thin trees, or a section of agricultural plants, thence continued other grass or green manure growing. Thick planting covers resist the force of raindrops falling directly to the ground, preventing the flow and soil from drifting down, creating favorable conditions for agricultural/sparse plants and fast growing/developing, so it has a great effect on both yield growth and protection. Intercropping improves the structure, the fertility, water permeability, and retention of the soil (Fig. 6.12)

- SALT model (Sloping Agriculture Land Technology) whose core is agro-forestry combination:

  - Hardware includes top forest stands with forester trees, fruit trees or other perennial trees and multi-purpose legumes row (acacia, pea, etc.) in contour planting to make green manure, animal feed, prevent erosion, keep moisture, create harmonious ecological conditions, and reduce pests.
  - The software includes various food crops, short-term foods, depending on the farmers' preferences, planted in the soil alternating between the double rows legumes.

SALT – A renewable agriculture on sloping land. Regenerative agriculture on sloping land is a practice aimed at improving slope land resources to increase land productivity and profit. Its distinctive feature is the promotion of the use of locally available, abundant resources and the reduction of outside investment (Source: Harold R. Watson, 1994).

Areas where a special attention should be paid to these techniques are the high-risk areas where many ethnic minorities live by agricultural production (Figs. 6.13 and 6.14)

Fig. 6.12 Planting under the anti-erosion tape

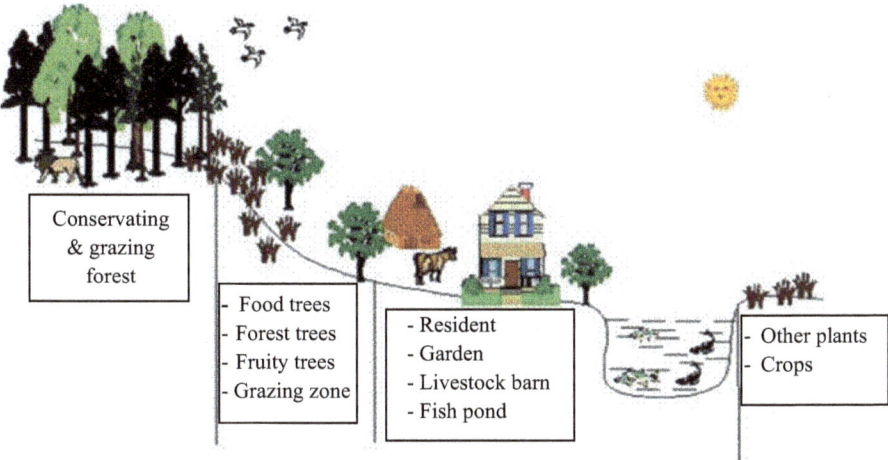

Fig. 6.13 SALT.1 model

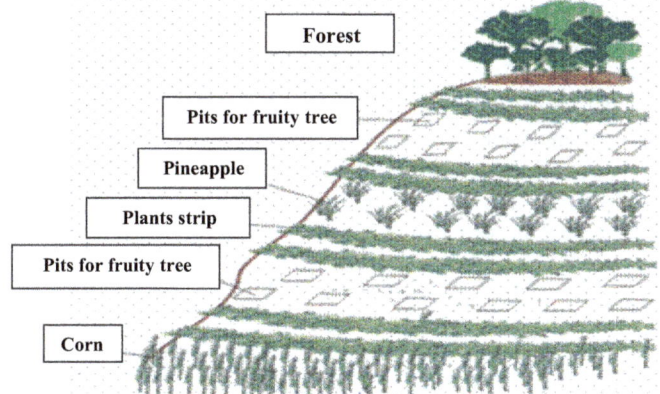

Fig. 6.14 SALT.2 model

- Phuoc Long district: Rat fountain.
- Dong Phu district: Cam fountain.
- Binh Long District: Fountains of Xa Cat, Heo, Rô, Vuon Khê swamp, Group 23 (Xa Cat fountain), Soc Siêu Lake, Villages of Nho, Trao, Bung, Bu Dinh, and Phung Lu.
- Loc Ninh district: K'Liêu fountain, Hamlets of 4, 6, Loc Binh, 10, 5b.

6.3 On Non-structural Solutions

Non-structural solutions should be given special attention, which allows eliminating and limiting the formation and movement of flash floods without causing any big changes to the basin environment, ensuring a sustainable development.

The objective of non-structural solutions is to influence the causes and mechanisms of flash floods formation and movement in order to eliminate or reduce flash floods. However, non-structural solutions such as management of basin and land use are not completely unrelated to the construction. Solutions such as warning, forecasting, evacuation from the threatening flash flood areas, and emergency relief are aimed at creating information about the basis to actively prevent and reduce damage when flash floods appear.

6.3.1 Land Use Management

Different from structural solutions for changing the flash flood process by dividing flood flows, preventing and storing flood waters, etc., managing land use for changing the composition and regulate land use on the basin areas, including land in major floodplains, to develop in the future to limit, or participate in the elimination of causes related to the buffer surface in the flash floods formation. Land use management usually includes two principled contents: control of land use assignment and exploitation, and control of construction and development.

Actually, the control of the exploitation, land use, and settlement in the province has not been consistently implemented. It can be seen that the formation of settlements near slopes, encroachments on beaches and banks of rivers and streams to build houses, obstruct flood drainage, creating additional causes of inundation when floods occur. Therefore, the land use planning must pay attention to the environmental impacts caused by flash floods:

– Restrict land exploitation in hilly areas, especially in districts of Bu Dang, Phuoc Long.
– Restrict the arrangement of residential areas along rivers and streams in town areas.
– The expansion planning for industrial and residential zones in the province requires the specific instructions to ensure safe prevention and reduce damage in water disasters and flash floods.
– Carry out production on low slope land along the belt road.
– It is necessary to have reasonable land use solutions, choose a reasonable cultivation method, not to cultivate on steep slopes of 25° angle or more. Less than 25° are cultivated, but erosion prevention must be addressed. Applying the advanced farming methods with agro-forestry models combined with afforestation and forest protection.

6.3.2 Land Schematic Planning in the Basin for Flood Prevention

The division of upper basins in the mountainous areas where heavy, flash floods often occur into zones depending on the hydrological characteristics is very important in finding the appropriate precautions, and to thereby have the right policy in construction, civil, commercial, and agricultural industries. A basin can be divided into zones where some keep the flow, some delay the flow, and some save the low area where floods threaten.

Water-catchment areas that functionally reduce flow (in upstream forests, streams, small rivers, including hills, mountains, terraces with developed forests capable of absorbing and storing rainwater) often do not allow developing the constructional forms. Especially avoid urbanization and all construction methods are at risk to decrease the natural function of flood reduction to preserve waters of the upstream. In this area, if the construction development is kept on a limited scale and there are measures to store rainwater to replace the natural role of the area, or to take the effective measures to control the flow.

In concave, low-lying areas and river banks that are always threatened by normal or flash floods, there must be a specific planning and design for each place depending on the level of inundation during floods. In particular, it is necessary to prohibit the construction of residential areas, houses on riverside, encroachment upon the conductor bed, even seal it there. Construction works, structural density, materials types, and installed machineries must have their technical characteristics considered under the impact of flash floods.

Using zonal schema planning in construction, it is possible to specify the flood indicators that designed for each area, for each work type, and to stand-by the availably supportive solutions to prevent and reduce damage. It is possible to do schematic planning for the purpose of construction and flood prevention according to the following steps:

- Schematic planning the basins and areas threatened by flash floods into the small zones according to hydraulic functions, as per flood frequency.
- Determine the land use method in each zone, thereby suggesting the construction policy and corresponding technical solutions.
- Determine the construction technical conditions and design criteria of the project according to the hydraulic function in each zone. Identifying solutions, technical measures, and supporting structures for avoidance.

6.3.3 Strengthening the Sustainable Land Management

It is necessary to have the programs and projects for research and implementation of long-term land use and management closely linked with socio-economic development programs at the macro and micro scale. There are long-term general research programs on soil fertility improvement and protection, combining the advanced technology transference with indigenous knowledge, ensuring sustainable land use, suitable for each region with different climatic conditions and cultivating techniques.

6.3.4   Afforestation and Forest Protection

To protect forests and plant forests in bare lands and bare hills to increase water permeability and water retention of the soil, and reduce surface runoff in case of heavy rain. Strengthen the protection of watershed forests, plant more trees to limit flash floods from turning into mud floods, muddy streams, migrating out of areas where flash floods often occur, with regular observation and monitoring in the rainy season.

6.3.4.1   Zoning for Reforestation

The zoning for reforestation is the cheapest and most effective method of forest building; however, it takes a long time. In the humid tropical conditions, there are also regenerated trees, and the reforestation zoning is a feasible solution. This measure is suitable in complicated terrain areas such as Binh Phuoc province. For this solution to succeed, a good protection must be implemented. It is forbidden to cut down, and when the forest has started to close the canopy, the forest will be nourished in a multi-layer application (Fig. 6.15).

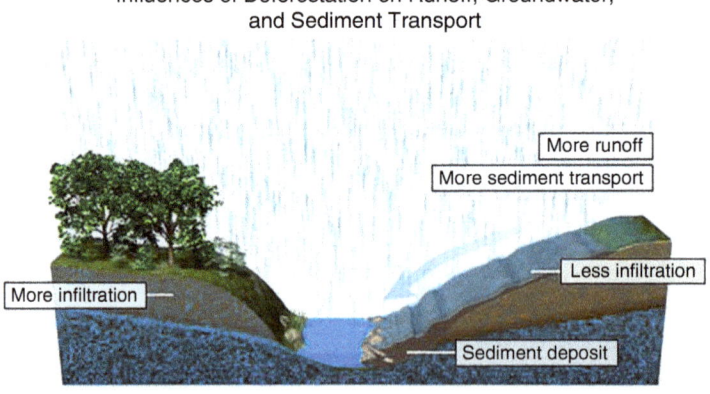

Influences of Deforestation on Runoff, Groundwater, and Sediment Transport

More runoff

More sediment transport

More infiltration

Less infiltration

Sediment deposit

©The COMET Program

Fig. 6.15 Bad impacts because of deforestation

6.3.4.2 Planting Protective Forests

Protective forest planting for flood prevention is both a measure of production and structure. With the main goal of protecting water sources against floods, the business goal and the method of planting protective forests are different from planting productive forests.

Principles of planting preventive trees for erosion and flooding: when planting forests to prevent erosion and floods in different areas, it is necessary to derive from natural geographic conditions, hydro-climatic conditions, and protection demand for suitable planting arrangements and selecting crops that fully meet the protection requirements.

However, because the main effects of forests are the regulation and control of flow, preventing soil swept away, sedimentation, when arranging for planting trees to prevent erosion and flooding, it is necessary to satisfy the following principles.

- Occupying adequate area enough to fill the role of water regulation for protecting land.
- Having the appropriate width enough to prevent the flow, maximize the effect of soil retention, sedimentation of mud and sand.
- The direction of the forest belt will be arranged along the contour line and prevent the entire flow.
- Forests against erosion and flood must have multi-layer structure, distributed on the most important places to strongly promote the capacity of water regulation and land retention.
- On bare land and slopes, sometimes a simply planted layer is good to shade and stimulate regeneration of mixed natural trees.

6.3.4.3 Principle of Selecting Forest Trees

For regulating water and preventing erosion on steep slopes

- The tree species must be suitable with ecological conditions of protection forest area.
- Must have broad, thick foliage, dense and evergreen branches to increase the foliage ability to resist rainwater, reduce the shock force of the raindrops into the soil.
- The tree has a deep and broad root system to fix the soil and convert many strong currents into underground currents.
- Plants grow fast, strongly, live perennially.
- The tree can withstand drought and poor bad soil.

6.3.4.4 Afforestation for Water Regulation on Slopes

To limit flood flow and the destruction of surface water flows, convert a part of surface flow into groundwater runoff, on agricultural production areas, fruit trees on slopes, and on upstream slopes where water concentration of the reservoir is; the forest strips should be planted to regulate the flow. When water flows through these forest strips, surface water flows will be dispersed, slow down, increase water permeability, and erosion of mud and sand will also be stopped.

6.3.4.5 Mixed Forest with Shrubs

This is a typical forest of basic flow regulating on sloping land. For the area of Binh Phuoc province, based on the factors of topography, soil, and climatic conditions, it is recommended to choose Acacia leaf, Concha Acacia, and Nacre tree For shrubs, it is possible to choose local plants such as Humble plant and Pineapple; mixed in rows, alternating trees with dense density.

6.3.4.6 Fruit Forest

Fruit forests are planted where the soil conditions permit high production (in the middle or at the foot of the slope). To promote the effect of water regulation, they can be mixed with 50% shrubs. If the soil is good with abundant fertilization, we can choose shrubs that also produce fruit or shrubs of great economic value for mixing. In Binh Phuoc province, it is possible to choose many species of fruit trees to mix such as Mango, Jack-fruit, Orange, and Lemon, Guava shrubs can be chosen with Pineapple and Downy Myrtle.

6.3.4.7 Shrub Forest

Shrub forests have a small capacity to regulate water but the ability to protect the soil is quite good. Therefore, if building a shrub forest with the goal of water regulation (flood control), the effect is not very high. However, the pure type of shrub forest is also suitable for relatively steep terrain conditions such as those in Binh Phuoc.

In the reclamation and use of hilly land for agricultural production and planting industrial trees, it is necessary to leave a part of the forest, the forest strip or the bush, natural grass on the slope to renovate the flow-regulating forest.

6.3.5 Works Protection, Conductor-Bed Clearance

This is a measure frequently used to reduce damage directly, drain flash floods faster, and avoid floods. Protecting works and clearing the conductor-bed are often associated with each other to avoid rivers and fountains, reducing damage.

Relocating of the structural works and communities out of areas impacted by flash floods can provide long-term economic, social, and environmental benefits, but often at high costs, also upset the normal life of the community and additional damage to services and trade. Therefore, this solution is only applicable in areas where severe flash floods are likely frequent occurrences.

6.3.6   Structural Works & Houses with Water-Proof Walls

This solution is widely used in residential zones, urban areas that are threatened with flash floods or inundation. The solution allows the use of the house walls themselves as artificial dykes to prevent and reduce the flood damage.

The design of the house walls in particular, the houses in general for this purpose requires additional calculations on the impact of flood flows, inundation, mud, and sand. The walls can be thin, but there may be a partition, between the bags of soil, sand, and clay to be tightly packed for solid resistance when necessary. This type of wall can be a temporary solution to prevent floodwater from overflowing through protected areas without affecting the arrangement of buildings and houses inherent on residential areas in the river valley. Of course, this typical solution is easy to mix with other types of flood protection structural measures once each house is designed in conjunction with other flood mitigation measures.

6.3.7   Flash Flood Zoning

Zoning for warning of flash floods danger. Zoning helps to identify areas at risk of variable severity, and from there, prepare the appropriate prevention. The zoning is usually realized on three levels: general, medium, and detail. This is the scientific basis for building the flood control plans (relocation, evacuation, rescue, relief, change of farming regime), for selection of specific structural and non-structural solutions for prevention; choosing to build forecasting and monitoring stations.

6.3.8   Prepare the Conditions for Disease Response

Preventing epidemics after flash floods is an important task. Local health organizations need to send a mobile team with specialist staff, drugs, anti-epidemic chemicals, and equipment, ready to give hospital assistance to localities regarding environmental sanitation and disease prevention during a flash flood.

Raise awareness for people in areas prone to flash floods. To avoid diseases, the most important thing issanitation, food, and drinking water.

6.3.9 Reasonable Construction Planning

- Warn local authorities about geologically critical locations that are prone to or will be damaged by flash floods, and do not plan settlements in those critical locations.
- No settlements on steep slopes, gorges, or gathering streams.
- Ensure that river and stream beds are well cleared so that floods can be drained easily. Do not build structures to obstruct flow. In case of building, it is necessary to arrange a system of culverts with appropriate aperture, ensuring flood drainage.
- For areas at risk of flash floods and mountain landslides, special attention should be paid to households living in low-lying areas along rivers and streams, in downstream of reservoirs, in mountainous slopes at risk of landslides that must resolutely direct the relocation through propaganda, mobilization, or coercion. In addition to immediately planning a new relocation to ensure safety for people after evacuating.

However, in some cases, the movement of an entire residential area with long-term living custom is difficult. Thus, in order to prevent and minimize the damage caused by flash floods in areas where people must live together, appropriate measures are required. One of those solutions is to schematically plan residential areas and exploit the economy accordingly. Principles of disposition of population zones and economic concentration regions are:

- Avoid the frequent inundation.
- Not affected by slope and mixed flash floods.
- Not situated in area of landslides and mud floods.
- Arranging the big structures to reduce flash flood intensity.

6.3.10 Coping with Flash Floods

6.3.10.1 Evacuate or Migrate Out of Areas Prone to Flash Floods

When there is a forecast or warning, the evacuation must be organized according to the plan already fixed. Therein, it is necessary to arrange different escape ways. Evacuation is the resettlement arrangement for people from likely dangerous areas due to imminent flash floods to safer areas. For evacuation to be effective, the early warning needs to be one step ahead.

This is a passive measure, only effective when it can predict the possibility of flash floods (information on warnings and forecasts of flash floods). Evacuation of people and property from the risky area, in fact, is a very difficult problem when people live scattered and lack means of transport, while information on flash floods is only known a few hours before (even later). However, this measure can only limit the damage in the valley where the floods pass, so evacuation is temporary, over short range and distance.

Evacuation can be considered as a short-term solution: it can be divided into three periods: before/during/and after flash flood. The pre-flood period is important in reducing damage, but is highly dependent on the expected timing of warnings and forecasts. During flash floods, often the evacuation process is mainly carried out when the flood is already happening. The post-flood period is mainly related to recovery from damage and restoration of areas in calamity.

In addition, it is necessary to develop a public education program to make the population aware of the migration plan and other responses to flash floods.

6.3.10.2 Search and Rescue

When flash floods occur, the advance-guard forces first must urgently rescue the victims and search for people to rescue.

6.3.10.3 Logistics and Supply

Provide minimal means of living and food to prevent hunger and disease during flash floods.

6.3.10.4 Information and Information Management

All of the above activities largely depend on information. Communication facilities such as radio, telephone, and support systems, including communication facilities and transmission lines, are required.

6.3.11 Overcoming the Consequences of Normal and Flash Floods

6.3.11.1 Overcoming and Settling

The main purpose of relief and recovery is to provide essential facilities and services to the community, families, and individuals to gradually stabilize and restore their normal life.

Resettlement and relocation of people to previous resident areas are also of great importance in overcoming consequences of natural disasters. In addition to overcoming with the material factors, the spiritual recovery for the community is the most important. This is a complicated and delicate work related to each community, each family, and each individual.

6.3.11.2   Damage Assessment

The effects of all emergency and prevention operations must be determined imme-diately after assessing the extent and total damage of flash floods. The detailed dam-age assessment provides a general framework for examining the damage separation by sector and cause. From there, it is possible to study and evaluate the preliminary effectiveness of the implemented preventive solutions, the coordination of actions in the prevention of the command agencies at all grades, re-evaluate the hazards of tubing and flash floods, and local inundation situation. This is the basis for correct-ing, supplementing, and perfecting the preventive measures.

6.4   Management Method

6.4.1   Flash Flood Warning and Forecast

When the weather, hydrology, and basin buffer surface shows the possibility of flash floods, the flood warning is immediately carried out. Flood warning and forecasting is considered as a special measure, very important among the non-structural preven-tive solutions.

Activities of flood forecasting and warning include:

– Data collection (both hydrometeorology and basin status).
– Data transmission.
– Meteorological forecast (weather forms, its activities, rain).
– Model, prediction plan.
– Prepare the warning.
– Transmission and dissemination of warning news.
– Receive warnings and take preventive actions.
– Reverse contact to adjust, correct the warnings.

6.4.1.1   The Typical Alerts

There are usually three types of alerts:

• The simplest first way, warning of flash floods when there is heavy rain in a small basin, the time of water gathering is very short with conventional flood forecast-ing techniques.
• Similar to the first type, warning when heavy, prolonged rains cover the basin upstream. For warnings, the information about rain on the basin should be evalu-ated as soon as possible.
• Using an electrical transmission system to send signals to alarm stations when flood flows can reach dangerous heights.

6.4.1.2   Flood Forecasts and Warnings

Special attention should be paid to the following stages:

- Data needed for forecasting and warning can include two types: data for designing and building the system and specialized data for the system to make warnings and forecasts. These two data types are very different. In which, the data part for designing the system construction must include data on the current status of the basin buffer surface, the network of rivers and streams, preventive works and other works.
- Data for the process of forecasting, warning, in fact is the development and testing of warning tools; forecasts, also include data on hydro-meteorological variables in the basin, on rivers, status characteristics in the basin (such as surface, slope, topography, soil, vegetation, erosion, conductor-bed features, status features of dangerous flood at locations). Data on rain causing flood, evaporation, permeability, and humidity are also used.
- When there are weather events affected by storms and tropical depressions, it is especially important to pay attention to the rainfall data before and during the expected time, which is especially important in forecasting.

6.4.2   Create the Policies on Flash Floods

The goal of this policy is to reduce and limit damage caused by normal and flash floods in both economic terms and flash flood intensity. Each locality needs to have its own policies to prevent flash floods in accordance with the specific characteristics of that locality.

Goals
- All flood damage preventions are intended to reduce hazards and the danger of destruction of life and property in the damaged area.
- Not causing increased hazards and future risks of damage in socio-economic development in flash-flooded areas. Flood-threatened lands must be planned and managed concerning the frequency and severity of flash floods.
- All warnings, forecasts, emergencies, and all the State's relief to mitigate and restore the territory in possible flash floods in the future. Information on past and upcoming floods must be fully provided to the people.
- Study the social, economic, and environmental results of flash floods and their impacts on individuals and the entire community.

  Other important contents of making a flash flood policy:

- Knowing the nature of flash floods in relation to other natural disasters such as earthquakes and landslides because flood policy must be included in the general policy on disaster prevention.

- Identify the agency or individual responsible for the selection of preliminary measures in drafting and implementing flood mitigation plans, for control and development in flood-prone areas.
- Sufficient policy, technical, and financial assistance to ensure reasonable progress is made in flood protection.

6.4.3 Develop and Complete a Comprehensive Flood Prevention and Management System

6.4.3.1 Co-ordination Measures

It is necessary to have close coordination of all agencies, units, and people who are exploiting the basin and resources such as land and water. Basin exploitation such as land use, urban development, and population growth will change the basin conditions, leading to structural changes, risk of damage, and increased damage there.

Comprehensive management of river basin covers all flood issues from urban drainage to inundation. The basin management approach from using land, water, plant sources to other sources is aimed at having a beneficial impact on the flood flow regime, flash floods, reducing the scale and extent of damage. The management of the basin depends closely on the possession of land and resources in the basin, the authority's ability to control activities for the basin flood prevention and mitigation. Strengthening the cooperation in basin management through regulations and laws allows for increased flood control efficiency.

6.4.3.2 General Management Solutions

- Do not reclaim to cultivate on steep slopes, especially where the slope is over 25°.
- Ensure that river and stream beds are well cleared so that floods can be drained easily.
- Be careful when there is heavy rain and pay attention to the flood warning bulletins broadcast on public media such as radio, television, and newspapers.
- Organize an elaborate examination, good review, and make statistics of residential areas, villages, hamlets, and households in areas that are directly affected by flash floods, mud and rock floods, landslides (riverside, streams, low-lying areas, slopes of roads systems, hillsides with thin surface soils layer are prone to landslides, downstream areas of reservoirs dams).
- On that basis, develop the detailed plans to prevent and avoid flash floods and landslides to communes, villages, hamlets, and households; Especially, carefully prepare plans on-place in remote residential areas and where prone to fragmentation during floods. There must be a plan to store the essential necessities such as food & alimentation, oil, emergency & epidemic medicine and must arrange support forces to handle possible bad situations.

- Immediately organize the warning signs staking in areas prone to flash floods and landslides, promptly notify and warn each household to proactively prevent and avoid. We must resolutely direct and organize the immediate relocation to a safe place for households living in areas at high risk of flash floods and landslides according to the alternate migration plan in the locality.
- Closely coordinate with the army stationed in the area, organize training for avant-garde forces in the area to raise awareness of flash flood disasters and check to complete the situation measures handling where natural disasters appear, instruct the test drills to raise the people's awareness & skills in proactive prevention and avoidance of casualties, and the ability to command and operate of the authorities and the Command Board for flood and storm prevention and control – Search & Rescue at all levels.
- Direct measures to strengthen management, protection, and restoration. Afforestation in areas prone to flash floods.
- Realize the integrative contents of flood prevention in building the population layout master plan, developing production, in building infrastructure and in directing implementation. First of all, pay attention to integrating these contents when implementing target programs of hunger eradication and poverty reduction; socio-economic development in extremely difficult communes; afforest 5 million ha of forests; and other programs in the area.
- To assign the popular media agencies to closely coordinate with specialized agencies to propagate and disseminate the knowledge and experience on flood prevention and natural disasters avoidance.

6.4.3.3   For Local Authorities

It is important to know: Because the majority of people in areas prone to floods often live there for a long time and in some places have never experienced floods, they are often subjective, when possible, so the evacuation warnings are often delayed and hesitant. Therefore, in urgent cases, coercive measures can be used.

Note
There are many different solutions to prevent and limit the harmful effects of tubing and flash floods. However, it must be based on each specific natural, social conditions and causes of zonal floods, to have a flexible coordination among solutions. For example, for places with high slope characteristics, measures should aim to separate flood flows, separate solid substances from flood flows, build control dams, and plant forests.

For areas with bare land and hills, land reclamation and afforestation must be used. Where prone to flash floods near residential areas, carry out evacuation, migration, and resettlement in safer places or build structures and houses with waterproof walls.

Some Conclusions

International experts believe that flash flooding in the mountainous and coastal areas of Vietnam and some other countries is a result of complicated weather conditions that may become a "new normal" form in the future. According to a global assessment of current meteorological conditions, Vietnam "is suffering from the worst climate change impacts in the world."

Climate change can affect the characteristics of rain and storms, and therefore the previous experiences may no longer be useful in the coming decades. In the future 10 years, damage from national disasters, flooding in first place, in Viet Nam could be at USD 4 billion if the government defers to take necessary and radical measures to avoid the loss of resources, environment, and power.

Therefore, researching flash floods and proposing solutions to prevent and limit their harm is very necessary. The research results will be an essential overview for everyone from lay people to researchers and managers. For the residents, they will know where and how dangerous they are, when disasters appear, such as landslides, flash floods, in order to decide whether to stay there or not, or if so, what to do in case.

For managers and researchers, they will have a panoramic view of flash floods, the risks of their occurrence frequency, and extent of their destruction to adjust the proper planning of residential areas, urban zones or migrate out of dangerous areas and have long-term and sustainable socio-economic development orientations, at the same time giving appropriate solutions to minimize the consequences of natural disasters.

Flash floods are a dangerous and unpredictable natural disaster. They often occur in mountainous and coastal areas, but now with the current momentum of mountain chiseling, road cutting, maximum urbanization, the rains and storms of the climate change period will drag flash floods to urban areas with numerous concrete architectural blocks to cause more disasters to life. With the disastrous power of nature, it is difficult to achieve the goal of effective prevention, but we must follow nature to respond, adapt, and take effective measures to maximally limit the hazards. This issue is only possible if the society is sustainably developed.

Nowadays, natural disasters are becoming increasingly severe, directly affecting the lives of many people around the world. According to scientists' estimation, if people continue to destroy nature as today, in the early twenty-second century, the sixth major extinction event will take place and people will face the totally of catastrophic destruction. Currently, humans cut down yearly about 15 billion trees, progressively more than 5000 billion trees have been destroyed by us for land. We are providing more lucrative opportunities to flash floods (Fig. 6.16).

More than half of the tropical rainforest is gone and half of the earth's arable land is now agricultural land, which means that very little fertile land remains for the wildlife. In the 2030s, the Amazonian forest area will be devastated, narrowed down to the point that it no longer has enough moisture to produce and store moisture. Mass forest fires will occur and the Amazon will slowly turn into a dead desert. The terrible flash floods and widespread Amazon barren forestland will turn South America into vast regions that no one dares to live in.

Fig. 6.16 Future green tree

The Earth's forest lung is disappearing; therefore, before the arrival of the twenty-second century, the global average temperature will be 4° C higher than today, making most of the land on the planet no longer able to support life. Then the sixth major extinction will happen, and humanity will perish. Throughout Earth's history, it has taken hundreds of millions of years for a global catastrophe to occur, and humanity only needed 200 years to do this. What will mankind do to prevent the extinction?

Addendum

Addendum 1 Vietnam typical flash floods in the past

No	Date	Affected area	Current	Rainfall	Loss	
					Human (dead/ injured)	Money (billion vnđ)
1	27/06/1990	Lai Châu Town	Nậm Lay stream	233	104/200	22
2	23/07/1994	Mường Lay, Lai Châu	Nậm Lay stream	187	34	18
3	17/08/1996	Lai Châu Province	–	258	89	21
4	3/10/2000	Nậm Coóng, Sin Hồ. Lai Châu	–	138	39	2
5	27/07/1991	Sơn La Town	Nậm La river	403	42	26
6	16/06/1990	Krông Bông, Giang Sơn, Lak, Đắc Lắc	4 minor reservoirs	370	22	3.4
7	29– 30/07/1999	Hàm Tân, Bình Thuận	Dinh river	300	27	187
8	19– 20/09/2002	Hương Sơn, Hà Tĩnh	Ngàn Phố river	500– 700	53/111	824
9	18– 19/07/2004	Du Già, Yên Minh, Hà Giang	–	200– 300	45/16	50
10	27– 28/09/2005	Văn Chấn, Yên Bái	–	233	50/8	162

Addendum 2 The zoning map/table/graph of flash flood risk in Bu Dang district of Binh Phuoc province is the basis for the layout of works for flash flood prevention in the district.

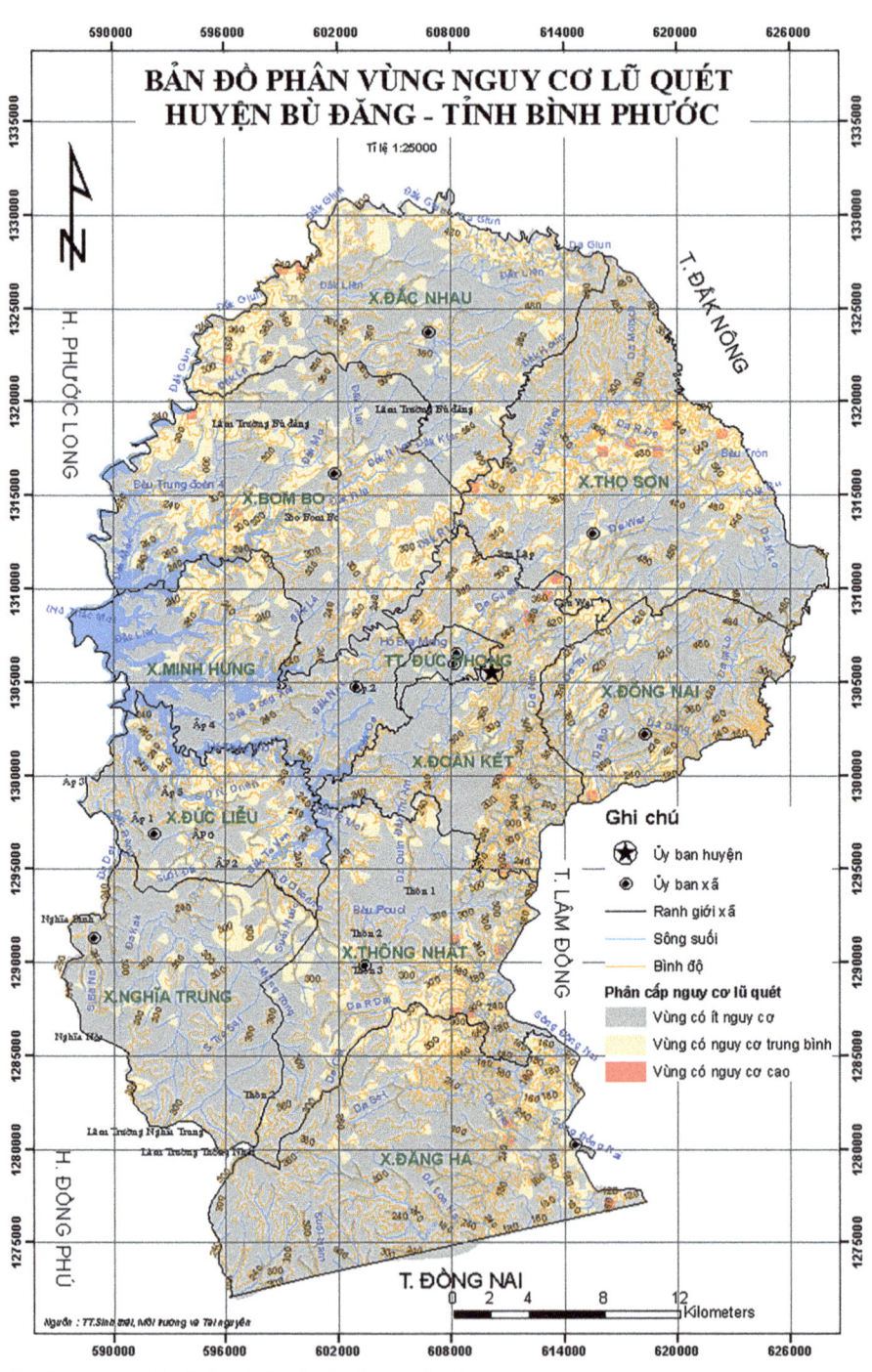

Zoning map of flash flood risk in Bu Dang district

Tables showing the percentage (%) of flash flood risk level in Bu Dang district

Code	Level of risk	Pixel quantity	Ratio (%)
1	Low	5089	80.72
2	Medium	1173	18.6
3	High	43	0.68

Graph showing the percentage (%) of flash flood risk level in Bu Dang district

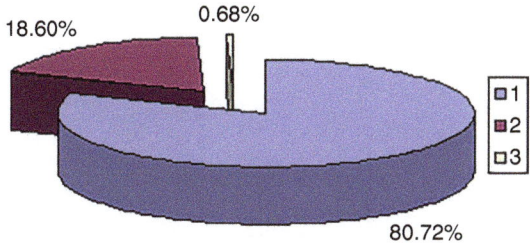

Comment The high risk ratio of flash flood, accounting for 0.68%, is distributed in the tributaries of Dak RLap and Dak Glun, Dăng Ha commune, hamlet 3 of Thong Nhat commune; Xuan Phu, hamlet 2 of Duc Phong town, Da Nao, Doan Ket commune, area along Dong Nai river. The average flash flood risk accounts for 18.6%, scattered in the Northwest of communes of Dak Nhau, Bom Bo, Son Thu, Son Lam, Son Hiep, Tho Son and the low risk of flash flood accounts for 80.72% in the Southern area of Tho Son commune, hamlet 6 of Duc Lieu commune, hamlet 3 of Thong Nhat commune, Thac Mo lake area and Northwest of Bu Dang district, East of Dang Ha commune.

Addendum 3 Zoning map of flash flood risk of Son Ha district, Quang Ngai province and of the whole Quang Ngai province, as the typical basis for the solutions of protective works against flash flood in the area.

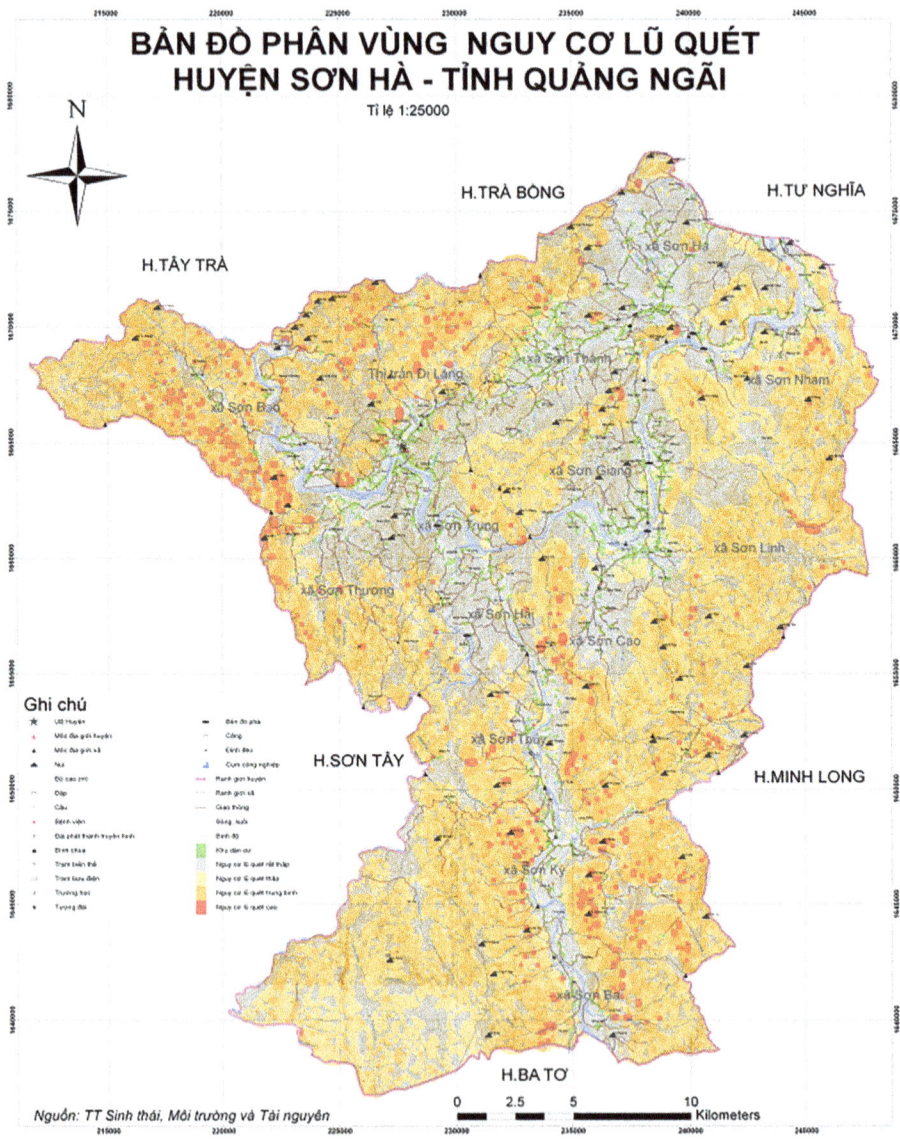

Zoning map of flash flood risk in Son Ha district of Quang Ngai province

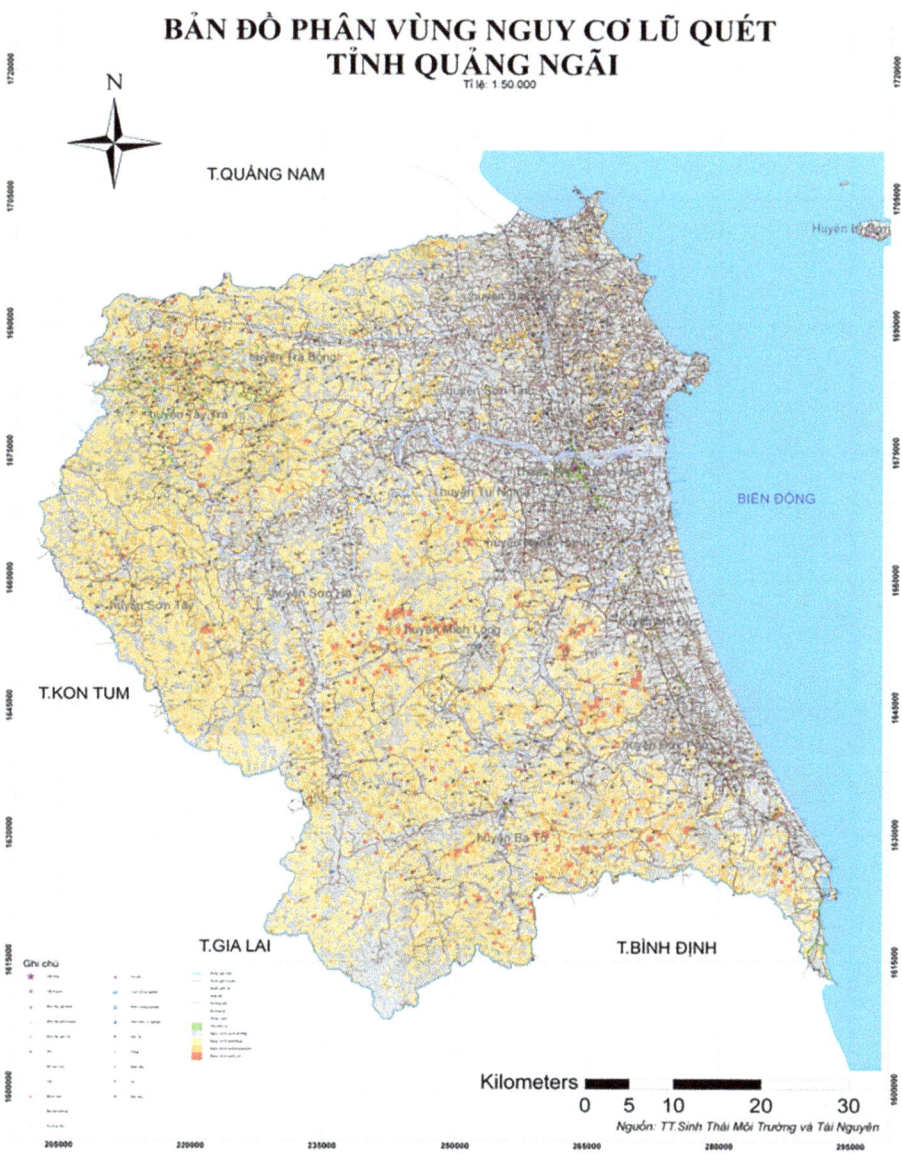

Zoning map of flash flood risk in Quang Ngai province

Glossary

Flash flood is a flood of short duration and high rainfall, when there is a storm, heavy rains gather quickly, causing floods on steep slopes. Flood waves can move very quickly causing sudden and serious damages. Because floods form in a short time, forecasting is difficult.

Digital Elevation Model (DEM) or Digital Terrain Model (DTM) is a digital model used to describe topographical elements, including geo-elevation, slope (gradient and direction), and slope types. DEM or DTM is one of the important applications in GIS that is the modeling capacity of topographic surface based on organizing data and methods in GIS.

Image interpretation using expert knowledge with the help of simple to complex optical instruments such as magnifying glasses, stereoscopic glasses, to identify the objects and factors.

Geographic Information System (GIS) is a collection of computer hardware, software, and procedures for storing, managing, controlling, analyzing, modeling, and displaying the geographic database toward solving the problems of management and complex planning.

Vector Paradigm is a "point-line-zone" model to describe the objects in the form of points, lines, and zones. The Descartes coordinate system (exp: x, y) and computer algorithms determine the coordinate points used in the vector system. The lines or arcs are a series of points.

Raster Paradigm based on a system of displaying, locating, and storing geographic data using the cell grid matrix. The coordinate shows of each pixel are its centroid. Each centroid or pixel in turn has its own data property assigned to them. The Raster data resolution depends on the pixel size or grid size, which can range from a few millimeters to many kilometers. Raster Paradigms are two-dimensional array data.

L. H. Ba et al., *Flash Floods in Vietnam*,
https://doi.org/10.1007/978-3-031-10532-6

Remote sensing (RS) is a technology to collect and analyze the information about any object without direct contact, by the electro-magnetic radiation (light, heat, microwave) or satellite/aerial photos to investigate and measure the object characteristics based on specific properties of reflection and emission.

LANDSAT ETM Photos (LANDSAT 7) LANDSAT is a US resource satellite managed by NASA. The Landsat remote sensing satellite system belongs to the resource satellite category operated by the EOSAT Earth Observation Satellite Company. To date, 7 generations of LANDSAT satellites have been researched and developed. LANDSAT 1 satellite was launched in 1972 with main document supplying sensor MSS. The new LANDSAT 7 satellite was recently launched into Earth orbit in April 1999 with an improved sensor TM (Thematic Mapper) called ETM (Enhanced Thematic Mapper), then improved to ETM$^+$ (Plus).

TM and ETM$^+$ TM is the photo made by mechanical scanning sensors that collect reflected and emitted energy in the visible field, reflected infrared between infrared and thermal infrared of the magnetic spectrum. It collects multi-spectral images with higher spatial and spectral resolution than Landsat MSS. The ETM$^+$ system has a panchromatic channel with a spatial resolution of 15 m and a thermal infrared channel with a spatial resolution of 60 m. The TM and ETM$^+$ channels were selected after many years of analysis for vegetation manifestations, soil and vegetation moisture measurements, differentiation of clouds and snow, and hydro-thermal variability in rocks.

The concept of plant index – NDVI The Normalized Difference Vegetation Index (NDVI) is used for evaluating if the plant coverage is high or low. NDVI is determined based on different plant reflections expressed between the visible spectrum, usually the red channel and the near infrared spectrum channel. Plant index NDVI is calculated using the formula. NDVI = (near-infrared channel – red channel)/(near-infrared channel + red channel) LANDSAT ETM photos have: The near infrared channel is channel 4 The red channel is channel 3 So: NDVI = (band 4 – band 3)/(band 4 + band 3) NDVI is in the range [–1,1]. NDVI has a negative value meaning that the visible channel has a higher reflectivity than that of the near-infrared channel, there is no vegetation.

Bibliography

Vietnamese Referred Documents

Le Huy Ba, Huynh Cong Luc. (2014) Flash Flood -theoretical basis and practice. (Agriculture Publishing House)

Le Huy Ba, Thai Le Nguyen, Huynh Cong Luc. (2016) Flash Flood -theoretical basis and practice in Vietnam (Science and Technology Publishing House)

Le Huy Ba, Nguyen Xuan Truong, Vu Ngoc Hung. (2018) Handling pollution, land degradation, riverbank and coastal erosion. Science and Technology Publishing House.

Nguyen Trong Yem and collaborators. (2005) Research to develop zoning map of natural environmental catastrophe in the territory of Vietnam. State level topic KC.08 by Professor.TS. Nguyen Trong Yem presides and taken over.

Lê Huy Bá, Huỳnh Công Lực et al. (2010–2013) Mapping the flash floods risk in Dak Lak province and proposing feasible solutions. (Final report of provincial scientific research project)

Lê Huy Bá, Nguyễn Thị Trốn, et al. (2003) Interaction of country environment and Mangrove West Ngoc Hien, Ca Mau (Final report of provincial scientific research project)

Lê Huy Bá, Nguyễn Văn Đệ (1993) Study on the possibility of soil erosion in the forest sub-basins of Vinh An, Vinh Cuu, Dong Nai Province. Branch subject under INDOFORUS Project (1991–1993), studying the possibility of soil erosion by basin in Southeast Asia (cooperation between Manchester University and universities of Vietnam, Malaysia, Indonesia, Thailand, funded by EC)

Lê Huy Bá, Thái Lê Nguyên and collaborators. (2008) Mapping the flash floods risk in Quang Ngai province and proposing feasible solutions. (Summary report on provincial scientific research projects 2005–2008)

Lê Huy Bá, Thái Lê Nguyên and collaborators. (2010) Mapping the flash floods risk in Binh Phuoc province and proposing feasible solutions. (Final Report of Provincial Project, 2009–2011)

Lê Huy Bá, Thái Văn Nam (2003) Evaluation of soil erosion in Dong Nai river basin. (Ministrial project, under the Ministry of Public Affairs 1999–2003)

Lê Huy Bá (2009) Land resources Environment of Vietnam (Vietnam Education Publisher)

Lê Huy Bá (1992) Descriptive study of laterite forms in Vietnm Southeast region. (Ministrial project 1990–1992)

Lê Huy Bá, Lê Văn Hòa, Nguyễn Thị Trốn (2005) Aquaculture ecological zoning in 8 coastal provinces of the Mekong Delta. (Report on the topic of the Ministry of Fisheries)

© The Editor(s) (if applicable) and The Author(s), under exclusive license to
Springer Nature Switzerland AG 2022
L. H. Ba et al., *Flash Floods in Vietnam*,
https://doi.org/10.1007/978-3-031-10532-6

Lê Ngọc Bích and co-authors (1995–1998) Investigate changes in the flow conductor-bed of the Mekong river system, downstream of Dong Nai - Saigon river and orienting the technical solutions to prevent landslides, reduce natural disasters on the Mekong River

Nguyễn Thế Biên (2001) The correlation between the direction of the shoreline and the erosion of the South Central Coast and the Southern -Collection of Scientific and Technological Results – Southern Institute of Irrigation Science, (Agriculture Publishing House)

Ministry of Science, Technology and Environment. (1995) Legal regulations of Environment. (National Political Publishing House, Hanoi, Vietnam)

Ministry of Agriculture and Rural Development (2002) Land assessment for land use planning and sustainable agricultural development in Dak Lak province. (Katholic University, Lueven -Belgium)

Ministry of Natural Resources and Environment -1/2006 – Circular guiding on implementation of contents of Decree No. 160/2005/ND-CP dated December 27, 2005 of the Government, detailing and guiding the implementation of the Law on Minerals and the Law amending and supplementing some Articles of the Law on Minerals

Ministry of Natural Resources and Environment -7/2006 – Regulations on decentralization of reserves and solid mineral resources.

Lê Thanh Bồn (2009) Lecture on soil science (Hue University of Agriculture and Forestry)

Tôn Thất Chiếu, Lê Thái Bạt, Nguyễn Khang, Nguyễn Văn Tân (1999) Handbook of land investigation, classification and evaluation, (Agriculture Publishing House)

Lê Thanh Chương (2012) Determining the cause of landslides on Tien River in An Long hamlet, An Binh commune, Long Ho district, Vinh Long province (Southern Institute of Irrigation Science)

Cao Đăng Dư, Lê Bắc Huỳnh (2000) Flash floods, causes and prevention measures – volume I and II, (Agriculture Publishing House)

Hydro-meteorological Station of the Central Region (2002) General hydro-climate of Quang Ngai province (Scientific report)

GIZ: Management of natural resources in coastal areas of Soc Trang province. Restoration of coastal areas and Mangroves with bamboo fences- Practical experience in Kien Giang province

GIZ, (2013) Restoration of coastal areas and Mangroves with Melaleuca fences Practical experience in Kien Giang province. (Australian AID cooperation program and Kien Giang Provincial People's Committee)

Lương Phương Hậu (2010) Topic KC08-14/06-10: Research on scientific and technological solutions for the system of river correction works in key sections of the Northern and Southern Deltas

Hoàng Văn Huân & collaborators (2000) Research on erosion control and coastal protection in Phuoc The, Tuy Phong, Binh Thuan province – Collection of scientific and technological results -Southern Institute of Irrigation Science, (Agriculture Publishing House)

Lê Mạnh Hùng, Đinh Công Sản (2002) Erosion of the Mekong River sides and prevention solutions for key areas, (Agriculture Publishing House, City. Ho Chi Minh City)

Lê Mạnh Hùng (2010–2013) Study on the influence of sand mining activities on the changes in Mekong River conductor-bed (Tien and Hau rivers) and propose management and planning of sand mining, (Southern Institute of Irrigation sciences)

Lê Mạnh Hùng, Trần Hoàng Bá (2008) Research & Assess the actual situation, determine the causes and propose solutions to prevent the landslides on riverbanks, canals, and flood control dikes in the Western districts of Tien Giang province" (Southern Institute of Irrigation sciences)

Lương Phương Hậu (1992) River Dynamics (Hanoi University of Construction)

Trịnh Thế Hiếu, Lê Phước Trình, Tô Quang Thịnh (2005) Current status and forecast of changes on coast and coastal estuaries in Vietnam

Lê Văn Khoa (2000) Soil and environment (Education Publishing House)

Nguyễn Viết Khoa, Đỗ Đại hải, Nguyễn Đức Thanh (2008) Cultivation techniques on sloping land, Ministry of Agriculture and Rural Development; National Agricultural Promotion Center, (Agriculture Publishing House, Hanoi)

Nguyễn Văn Mạo, Nguyễn Đăng Hưng (2004) Research on new technology, analysis of erosion causes and preventive solutions for Binh Thuan province (Science and Technology Journal of Irrigation and Environment)

Nguyễn Quang Mỹ (1979) Feature of the East-South Central Highlands. (Journal of Scientific Activities)

Nguyễn Thanh Phương, Trương Công Cường, Actual status of cultivation on sloping land in Dak Nong province, (Institute of Agricultural Science and Technology South Central Coast)

Nguyễn Tử Siêm, Thái Phiên (1999) Vietnam hilly land – Degradation and recovery, (Agriculture Publishing House)

Phạm Trọng Thịnh, Hoàng Thơi, Trần Huy Mạnh, Lê Trọng Hải, và Klaus Schmitt, (2009) Technical guide on Mangrove restoration and management. (German Technical Cooperation Organization)

Trần Văn Tư (2006a) Actual status and planning direction of areas in frequent flash floods and landslides (Institute of Geology, Vietnam)

Nguyễn Đình Vượng (2010) Assess the process of coastal erosion in Binh Thuan province, analyze the causes and propose the preventive solutions (Southern Institute of Science and Technology)

Nguyễn Vy, Đỗ Đình Thuận (1976) Main types of soil in Vietnam (Science and Technology Publishing House)

Central Geological Federation (2006) Synthesize the geological and mineral maps of Quang Ngai province; Proposing solutions for investment in exploration, exploitation and use of key mineral resources (Report on results of scientific projects)

Nguyễn Quang Mỹ (2005) Modern soil erosion and anti-erosion measures, (Hanoi National University Publisher)

Nguyễn Quang Mỹ, Quách Cao Yêm, Hồng Xuân Cơ (1984) Erosion research and test of some measures to prevent erosion of agricultural land in the Central Highlands (State Committee of Science and Technology -Ministerial-level project of the Central Highlands Integrated Research Program 1976–1980)

Thái Phiên, Nguyễn Tử Siêm (1998) Cultivation to protect sloping land in Vietnam (Agriculture Publishing House)

Thái Phiên, Nguyễn Tử Siêm (2002) Sustainable use of mountainous and highland land in Vietnam (National Publishing House, Hanoi)

Department of Natural Resources and Environment of Dak Nong Province (2009) Report on the current status of Environment in 2009 of Dak Nong Province

Department of Natural Resources and Environment of Binh Phuoc Province (2006) Report on the current status of Environment in 2006 of Binh Phuoc Province

Department of Natural Resources and Environment of Quang Ngai Province (2005) Report on the current status of Environment in 2005 of Quang Ngai Province

Actual status of coastal erosion – accretion in Binh Thuan province. (Journal of earth sciences)

Center of Environmental Technics (2009) *Assessment of impacts due to sedimentation, landslides*. Project "Determining methods to predict landslides and establish a stable riverbank corridor for economic development -society in Ben Tre province" (Department of Industry and Trade of Ben Tre province, 2007)

Interdisciplinary Center for Remote Sensing and GIS – Ministry of Agriculture and Rural Development (2000) Scientific project report: "Evaluation of Erosion Potential in the North Central Highlands of Vietnam"

Trần Văn Tư (2006b) Current status and planning direction of areas with frequent flash floods and landslides (Institute of Geology. Vietnam Academy of Science and Technology)

Institute of Soil Chemistry (2001) Basic information on the main types of soil in Vietnam, World Publishing House

Foreign Documents Referred

Blaney D, Warrington, G. (1983) Estimating Soil Erosion Using an Erosion Bridge. USDA Forest Service, Watershed System Development group: Fort Collins, CO

Franklin & Hampden (2003) Massachusetts erosion and sediment control guidelines for urban and suburban areas

Pitt R (2004 Alabama) Module 4: Erosion Mechanisms and the Revised Universal Soil Loss Equation (RUSLE). University of Alabama

Bryan Lee C, Lorch Z and Melong T (2012) Evaluation of the erosion control methods implemented by the panama canal expansion program

Hartman J (2001) Combination of technologies improves costal erosion measurements

Health NDD (2001). A guide to temporary erosion control measures for contractors, designers and inspectors

Jin GEJA (2010) A field study on cost-effectiveness of five erosion control measures. Management of Environmental Quality: An International Journal

Jones DK (2009) Calculating Revised Universal Soil Loss Equation (RUSLE) estimates on department of defense lands: a review of rusle factors and U.S. Army Land Condition-Trend Analysis (LCTA) Data Gaps

Meyer LWW (1969) Mathematical simulation of the process of soil erosion by water. Transactions of the American Society of Agricultural Engineers

Morgan RPC (1996a) Soi erosion and conservation. Longman

Abbott PL *Natural disasters*. San Diego State University

Morgan RPC (1996b). Soil erosion and conservation. Longman Limited

Lal R (1990) Soil erosion in the tropic. McGaraw-Hill, Inc

Rickson R (2006). Controlling sediment at source: an evaluation of erosion control geotextiles. Earth Surface Processes and Landforms. Stone, R. (Ontario, 2000.).

Roose EJ (1977). Use of the universal soil erosion equation to predict erosion in West Africa. In soil erosion: prediction and control. Soil Cons. Soc. Am., Ankeny, Iowa, pp 60–74

Morrow R (1994) Guide on land use according to sustainable agriculture. Agriculture Publishing House

Universal Soil Loss Equation. Ontario Ministry of Agriculture, Food and Rural Affairs.Wischmeier W.H. and Smith D.D (1978). Predicting Rainfall Erosion Losses, USDA Agr. Res. Serv. Handbook

Wischmeier WH, Smith DD (1978) Predicting rainfall erosion losses, a guide to conservation planning. Agriculture Handbook, No. 537, US Department of agriculture, Washington DC

United Nations Environmental Programme (1998) Training Guide on Environmental Impact Assessment (Hà Nội, Vietnam).

Asst Professor Binata Roy and Dr. AKM Saiful Islam [Institute of Water and Flood Management (IWFM), Bangladesh University of Engineering & Technology (BUET). (In Search of a Real-time Flood Forecasting System – Article on **eDailySun** magazine -July, 2020)

Index

Milton Keynes UK
Ingram Content Group UK Ltd.
UKHW020112280923
429512UK00001B/26

9 783031 105340